Primary Processing and Value Addition and Cold Chain Logistics

CONTENT VIBES
PUBLISHERS • PRINTERS • AGGREGATORS

About the Author

Dr. Sarvesh Kumar was born at Etawah, Uttar Pradesh (India). He did his B.Sc. (Ag) Hons & M.Sc. (Ag) from CSAUA&T Kanpur, UP during 2001-2007. He received his Ph.D. in Extension Education from BHU, Varanasi (UP). He faced ARS twice in 2009, and 2010 and qualified ASRB (NET) respectively. Presently he is working as an Associate Professor, the Department of Extension Education, Institute of Agricultural Sciences, BHU, Varanasi (UP), India. He has worked as a Scientist (Agricultural Extension & Rural Sociology) for 12+ years (2011 to 2024) at Jawaharlal Nehru Krishi Vishwavidyalay, Jabalpur (MP) which has merit of the best Agricultural University in India (2018). He is the recipient of a Book Prize & Certificate from CSAU&T, Kanpur for being a topper of the Department of Agricultural Extension, M.Sc. (Ag) batch 2007. He has also received the Mandi Parishad Merit Scholarship during his B.Sc. (Ag) Hon's & M.Sc. (Ag) and the UGC Research Scholarship during his PhD at BHU, Varanasi (UP). He was designated as a member of the Anti-ragging Committee and elected representative of Research Scholars of the Institute of Agricultural Sciences, BHU, Varanasi (UP), India during 2007-2010. The excellent extension, teaching and research work done and many times was recognized by his host institution and district collector including an appreciation letter given by the Hon'ble Vice Chancellor, Agricultural minister, Minster to women and child developed Govt. of MP, and other national & International societies. His teaching research temperament may be witnessed by his research publication in nut shell as NAAS rated research papers 44, books 05, book chapters 12, extension bulletins-12, folder-40, articles English-20, articles hindi-131, ICT developed -03, varieties registration assistance under PPV&FRA 09, teaching-RAWE program & AIA-12 years, total award received-16, total appreciations & recognition-16, FPO developed -05, Common service centers developed-05, value chain development-10, HRD training received for 98 days, successful case & Impact registered by self-15, sponsored training given 239 days, In-service training-12, OFTs-12, FLDs-07 organized, training for farmers & farm women-538, training for rural youth-16, field days-48, Mela exhibitions-46, recognized fellowship received-03, invited lecturer given-554, paper presented national/international level-33.

Dr. Neeraj Singh is an alumnus of B.H.U. and School of Management Sciences, Varanasi, as he graduated from the Faculty of Commerce, B.H.U., in 2008 and PGDM in 2010. Dr. Neeraj obtained his PhD degree from the Institute of Management Studies, BHU in 2023. Dr. Singh has a rich +12 years of teaching and 2 years of corporate experience. Currently engaged as an Assistant Professor in MBA (Food and Agribusiness) formerly MABM, in the Department of Agricultural Economics, Institute of Agricultural Sciences, Banaras Hindu University. Dr. Neeraj has a keen interest in the fields of Human Resource Management, Communication, Agribusiness Environment and Policy, Strategic Management, Business Law and Ethics.

He has published more than fifteen research papers in international and national journals, 1 patent and 5 book chapters to his credit. He has attended 35+ conferences, and seminars and presented more than 29 papers at the event, and has also delivered lectures in winter schools organized by the department. He has also attended numerous FDP of one-week and three-week durations.

Primary Processing and Value Addition and Cold Chain Logistics

Sarvesh Kumar
Associate Professor
Department of Extension Education
Institute of Agricultural Sciences
Banaras Hindu University
Varanasi-221005, UP

Neeraj Singh
Assistant Professor
Department of Agricultural Economics
Institute of Agricultural Sciences
Banaras Hindu University
Varanasi-221005, UP

CONTENT VIBES
PUBLISHERS • PRINTERS • AGGREGATORS

CONTENT VIBES
PUBLISHERS • PRINTERS • AGGREGATORS

1168, Sector 13, Urban Estate
Karnal-132 001, Haryana
Tel: 91-84470 75807, 18440 41168
Email: contentvibesppa@gmail.com
www.contentvibes.in

Print ISBN: 978-93-67559-82-6

ebooks ISBN: 978-93-67556-96-2

Preface

The world of agriculture is constantly evolving, with new technologies, methods, and strategies shaping the industry. Among the most significant advancements in recent years are those related to primary processing, value addition, and cold chain logistics. These aspects play a crucial role in ensuring that agricultural products are not only preserved but also transformed into higher-value goods that meet the ever-growing demands of global markets. The importance of these processes cannot be overstated, as they contribute directly to reducing food wastage, enhancing product quality, and improving the income of farmers and agribusinesses.

This book, Primary Processing and Value Addition and Cold Chain Logistics, delves into these essential components of the agricultural value chain. It explores the various techniques and technologies involved in transforming raw agricultural products into finished goods that are suitable for consumption or further processing. Additionally, it highlights the critical role of cold chain logistics in maintaining the quality and safety of perishable products as they move through the supply chain, from farm to market.

Throughout the chapters, we examine the economic, environmental, and social benefits that arise from effective primary processing and value addition. We also shed light on the challenges faced by the industry, such as the need for infrastructure development, technology adoption, and regulatory compliance. By providing a comprehensive understanding of these processes, this book aims to offer valuable insights to farmers, agribusinesses, policymakers, and researchers alike.

As the agricultural sector continues to grow and evolve, it is crucial to embrace innovations in primary processing, value addition, and cold chain logistics. This book serves as a guide to navigating this dynamic landscape and unlocking the full potential of agricultural products for both local and global markets.

—Author

Contents

1

Primary Processing of Fruits and Vegetables

Primary processing of fruits and vegetables refers to the first steps in transforming raw agricultural products into forms more suitable for consumption, storage or further processing. This step generally includes cleaning, sorting, grading, cutting, peeling and packaging of the products. The aim of primary processing is to preserve the nutritional quality, taste and appearance of the products by extending their shelf life and making them more suitable for consumers or ready for secondary processing. For example, fruits may be peeled, cut or peeled, and vegetables may be washed, chopped, or blanched. This step is essential in the agribusiness industry because it has a direct impact on the quality and safety of the final product, ensuring that it meets market requirements, food safety standards and consumer preferences. The efficiency and methods used in primary processing significantly affect overall profitability, product quality and market competitiveness.

Primary processing refers to the initial stages of preparing fruits and vegetables for consumption while preserving their nutritional value, freshness and flavour. This essential step involves various techniques for cleaning, sorting and packaging the produce, ensuring that it reaches consumers in the best possible condition.

SORTING AND GRADING OF FRUITS

Selection of Fruits and Vegetables

- Fruits and vegetables must be absolutely fresh.
- Fruits should be ripe, but firm and uniform. Ripe fruits should be rejected because they are infected with microorganisms and result in a poor quality product. Unripe fruits should be discarded because they usually shrink and harden during storage.
- All vegetables, except tomatoes, should be soft.
- The tomato must be firm, ripe and dark red in colour.
- The fruit and vegetables must be free from impurities.
- They must be free from blemishes, insect damage or mechanical injuries.

Grading of Fruits

The selected fruits and vegetable are graded by size and color to achieve uniform quality. This operation is done manually or by machinery such as screw graders and roller graders. Fruits such as berries, plums and cherries are graded whole, while peaches, pears, apricots, mangoes, pineapples, etc, are usually graded after being cut into pieces or slices.

In India, manual grading of fruits according to their size is generally done. The growers usually employ skilled packers who also grade the fruit based on their experience.

- Few mechanical sizing machines have been developed in the country like stretch belt sizing, mechanized grading, washing, waxing and grading units for Nagpur mandarin, mosambi and kagzi lime.
- Mechanical grader for black pepper and raw cashew nuts.
- Onion halving and fruit spherical cutting machine.
- Mango grader and Sapote grader.
- All these graders should be popularized among farmers after considering accessibility.
- Development of a complete packaging line with grading machinery, dehydration unit, hot water treatment unit, washing, waxing, dehydration and packaging for mangoes can be used to reduce losses and ensure good yields.

WASHING OF FRUITS AND VEGETABLES

It is important to remove pesticide residue and dust from vegetables and fruits. One gram of soil contains 10^{12} spores of microorganisms. Therefore, it is essential to eliminate microorganisms by washing with water. Fruits and vegetables can be washed in various ways. Roots that are exposed to the soil are washed with water containing 25 to 50 ppm of chlorine (as a detergent). Other washing methods are spray washing, steam washing, etc.

RIPENING OF FRUITS

For some fruits collected unripe, such as avocado, banana, kiwi, mango, nectarine, papaya, peach, pear, plum, melon, etc., it may be necessary to ripen before processing. Ethylene treatment can be used to achieve faster and more uniform ripening. The optimum temperature for cooking is 15-25° C. The higher the temperature, the faster the ripening. The relative humidity should be maintained between 90 and 95% during cooking. A concentration of about 10 ppm of ethylene in a closed room is sufficient to promote ripening. Ethylene is produced by the reaction between calcium carbide and moisture, mainly those used in the fruit trade to accelerate the ripening process. However, indiscriminate application can pose serious health risks. Ethephon is used commercially for the pre-harvest ripening of fruits, berries, tomatoes and coffee. It is also used to facilitate the picking of fruits and berries (to loosen the fruit) and to accelerate ripening after harvest. It is essentially a plant growth regulator with systemic properties. It penetrates into tissues and is transported. It breaks down into ethylene, which is the active metabolite.

Despite the fact that artificial ripening methods like ethereal treatments can help ensure uniform ripening of fruits such as bananas and sapotas, there are still significant challenges in implementing these methods on a large scale. While ethereal and other chemical treatments (like 2, 3, 5-TP and maleic hydrazide) are effective, particularly for bananas and sapotas, the use of banned chemicals like calcium carbide remains a concern in many regions.

Additionally, the lack of affordable and accessible ripening chambers for small-scale growers limits their ability to adopt more regulated, uniform ripening practices. Most large-scale enterprises have the resources to set up these chambers, but smaller producers often do not have access to the necessary infrastructure. Therefore, while there are promising alternatives to harmful chemicals, the gap between large enterprises and small growers continues to affect the widespread adoption of safer, more uniform ripening methods.

Blanching of Fruits and Vegetables

Blanching involves briefly immersing fruits or vegetables in boiling water and then quickly cooling them. This process helps preserve color, flavor, and nutrients by inactivating enzymes that could cause spoilage. Blanching is commonly used to freeze produce, allowing consumers to enjoy out-of-season fruits and vegetables without compromising their nutritional value. Also called blanching, steaming, or pre-cooking.

Fruits are generally not blanched, which leaves the oxidative enzyme system active. Sometimes, the fruit is soaked for a period of time—from half a minute to five minutes, depending on the variety—in water at a temperature between 180°F and 200°F, then cooled immediately by immersion in cold water. The purpose is to loosen the structure and thus allow more weight to be pressed into the container without damaging the individual fruits. Blanching is usually done in the case of vegetables by exposing them to boiling water or steam for 2 to 5 minutes, followed by cooling. The degree of blanching varies among varieties. This brief heat treatment involves the following tasks:

- Deactivates most plant enzymes responsible for issues like toughness, discoloration (such as polyphenol oxidase), mustiness, off-flavors (like peroxidase), softening, and nutrient loss.
- Reduces the volume of leafy vegetables like spinach by causing shrinkage or wilting, making them easier to pack.
- Removes tissue gases that contribute to the reduction of sulfides.
- Reduce the number of microorganisms by up to 99%.
- Improves the green color of vegetables such as peas, broccoli, and spinach.
- Removes saponins from peas.
- Eliminates undesirable acids and astringency in the peel, improving the flavor.
- Assists in peeling vegetables like beetroot and tomatoes by removing their skin.

REDUCTION OF SIZE

Size reduction refers to the process of breaking down large solid raw material units into smaller fragments or particles. This is achieved by various mechanical means such as cutting, grinding, crushing and abrasion. The main purpose is to increase the surface area of the material, which facilitates subsequent processing steps such as mixing, cooking or chemical reactions.

Grinding Equipment

Grinding equipments categorized into two types: (1) crushers, and (2) grinders. Crushers primarily rely on compressive forces, while grinders use a combination of shear, impact, and compressive forces.

Crushers

Jaw and gyratory crushers are large, heavy-duty machines that are rarely used in the food industry. In a jaw crusher, material is placed between two sturdy jaws—one stationary and the other moving back and forth. This motion gradually forces the material into a smaller space, crushing it as it moves downward.

A gyratory crusher, on the other hand, consists of a conical chamber with a rotating crushing head inside. The head moves in an eccentric manner, pressing the material between an outer fixed cone and an inner gyrating cone. As the material moves downward, it is continuously compressed into a tighter space, breaking apart in the process. Both crushers operate by applying pressure to reduce material size, as shown in Fig. (a) and (b).

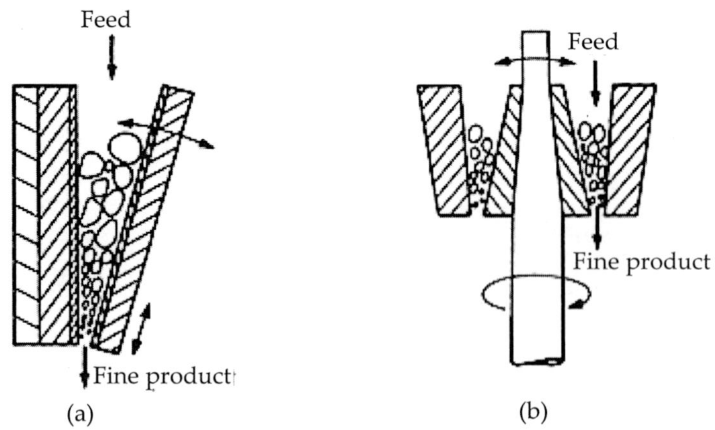

Figure: 1.1 Crushers: (a) Jaw, (b) Gyratory

Crushing rolls are composed of two sturdy, horizontally aligned cylinders positioned parallel and in close proximity to each other. These cylinders rotate in opposite directions, capturing and compressing the material as it moves through the narrow gap between them. In certain setups, both rolls operate at the same speed, while in others, they may rotate at different speeds, or only one roll may be powered. A key use of this mechanism is in the cane sugar industry, where multiple sets of rolls are employed to extract juice by crushing the sugarcane in successive stages.

Hammer Mills

In a hammer mill, swinging hammerheads are attached to a rotor that rotates at high speed inside a hardened casing. The principle is illustrated in Fig. 1.2 (a).

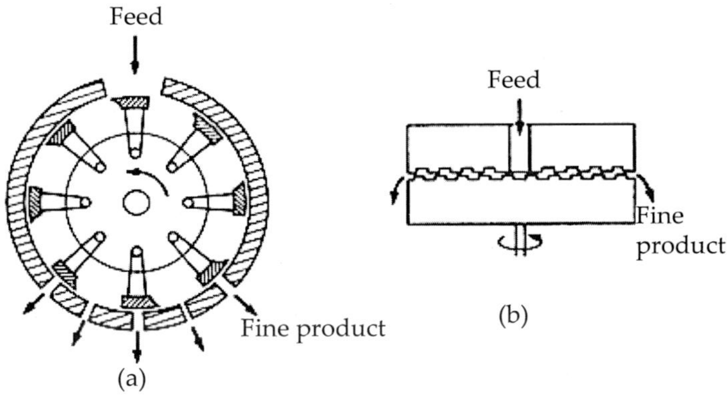

Figure: 1.2 Grinders: (a) Hammer Mill, (b) Plate Mill

The material is broken down and ground between the hammers and the casing, staying inside the mill until it becomes fine enough to pass through a screen at the bottom. Hammer mills can process both brittle and fibrous materials, though for fibrous substances, specially designed projections on the casing may be added to enhance the cutting action.

Plate Mills

In plate mills, the material is placed between two circular plates, one fixed and the other rotating. The feed is introduced near the center of the plates, where it is sheared and crushed as it moves toward the edges, see Fig. (b). These mills can have the plates mounted horizontally, as seen in traditional Buhr stones used for grinding corn, which feature fluted surfaces. Alternatively, the plates can also be mounted vertically. Advances in plate mill technology have led to the development of colloid mills, which use extremely fine clearances and high speeds to produce particles of colloidal size.

Fixed Head Mills

Different types of mills are used where the material is sheared between a fixed casing and a rotating head, typically with very small clearances between them. One example is the pin mill, where both the stationary and moving plates have pins attached to their surfaces, and the powder is sheared between these pins.

Roller Mills

Roller mills function similarly to roller crushers but feature smooth or finely grooved rolls that rotate at different speeds. They are extensively used for flour milling. Due to their straightforward design, the maximum particle size that can pass between the rolls can be controlled. If the friction coefficient between the rolls and the feed material is determined, the largest particle that can be captured between the rolls can be calculated based on the particle geometry.

Miscellaneous Milling Equipment

Milling equipment comes in a wide variety of types. One example is the ball mill, where the material to be ground is placed inside a horizontal cylinder or cone, and then tumbled with steel balls, natural pebbles, or artificial stones to crush and break the material. While ball mills have limited uses in the food industry, they are employed for grinding food coloring materials.

Another type is the edge runner mill, which features a heavy, broad wheel that rotates around a circular trough, commonly used for grinding chocolate and confectionery.

Many milling machines have become standard in various industries, with some claiming that their specific actions are uniquely suited to, and necessary for, producing particular products.

IMPORTANCE OF SIZE REDUCTION IN FOOD PROCESSING

Size reduction is an essential aspect of food processing due to several key benefits .

- **Better Mixing and Blending:** Reduced particle size leads to more uniform mixing, ensuring consistent product quality.
- **Improved Cooking Efficiency:** A larger surface area facilitates faster and more even cooking.
- **Enhanced Texture and Appearance:** Finer particles contribute to a smoother consistency and a visually appealing final product.
- **Higher Extraction Yields:** Smaller particles improve the efficiency of extracting oils, flavors, and other valuable components.
- **Greater Energy Efficiency:** Properly sized particles minimize the energy needed for further processing stages.

Principles of Size Reduction

Size reduction in food processing is based on four main principles: (i) compression, (ii) impact, (iii) attrition, and (iv) cutting.

Each principle is designed for specific types of materials and applications.

Compression

Compression is the process of applying pressure to a material to reduce it into smaller fragments. This method is typically used for materials that are brittle or have a crystalline structure, such as grains and seeds. Equipment like roller mills and jaw crushers are commonly used to achieve compression.

Key points:

- Best for brittle materials
- **Equipments:** roller mills, jaw crushers
- **Applications:** grain milling, seed crushing

Impact

Impact is the act of striking a material with force to break it into smaller pieces. This principle is ideal for hard or fibrous materials, such as fruits, vegetables, and meat. Impact can be achieved using equipment such as hammer mills and impact crushers.

Key Points:

- Best for tough or fibrous materials
- **Equipments:** hammer mills, impact crushers
- **Applications:** fruit and vegetable processing, meat processing

Attrition

Attrition is the process of breaking down a material through friction and shear forces. It is particularly effective for materials that are abrasive or have high moisture levels, such as spices and moist grains. This can be done using machines like attrition mills and disc mills.

Key Points:

- Best for abrasive or high-moisture materials
- **Equipments:** attrition mills, disc mills
- **Applications:** spice grinding, wet grain milling

Cutting

Cutting is the process of dividing a material into smaller pieces using sharp blades. This method works best for soft or elastic materials, such as fruits, vegetables, and dough. It can be done using equipment like slicers, dicers, and cutters.

Key Points:

- Best for soft or elastic materials
- **Equipments:** slicers, dicers, cutters
- **Applications:** fruit and vegetable slicing, dough cutting

Selecting the Right Equipment for Size Reduction

Selecting the right equipment for size reduction depends on various factors, including the material type, desired particle size, and production requirements. Key factors to consider include the following:

- **Material Properties:** Assess the hardness, moisture content, and abrasiveness of the material.

- **Target Particle Size:** Define the desired particle size and distribution for the specific needs.
- **Production Capacity:** Ensure that the machinery can manage the necessary processing rate and throughput.
- **Energy Efficiency:** Choose equipment that minimizes energy use while achieving the desired reduction in size.
- **Maintenance and Durability:** Select equipment that is easy to maintain and durable enough to meet the demands of the processing environment.

Optimizing Size Reduction Processes

To attain the best possible size reduction, it's important to fine-tune your processes and equipment settings. Below are some tips to help optimize size reduction in food processing:

- **Monitor Particle Size Distribution:** Regularly check the distribution to ensure consistent product quality, adjusting as necessary.
- **Adjust Equipment Settings:** Fine-tune settings such as speed, pressure, and blade sharpness to reach the desired particle size while minimizing energy use.
- **Implement Process Controls:** Use sensors and automation to monitor and adjust size reduction parameters in real-time.
- **Maintain Equipment:** Conduct regular inspections and maintenance to prevent wear, ensuring consistent performance and reducing downtime.
- **Train Operators:** Provide training to operators so they understand the principles of size reduction and can make informed adjustments to the process.

Effect of Size Reduction

Advantages

Size reduction offers numerous benefits. It allows food to be processed into a more convenient, preferred, or necessary size. For instance, reducing the bulk density of materials makes transportation easier and more cost-effective. It also improves mixing efficiency, heat transfer, and the extraction of food components like flavors and oils. From a consumer perspective, it enhances sensory qualities, particularly texture and appearance. For example, finely milled corn is often preferred for porridge over coarser grits, and turkey slices are typically favored in sandwiches over minced turkey.

Disadvantages

Size reduction, such as milling, can result in the loss of volatiles, especially in heat-sensitive foods like spices. In wheat milling, valuable nutrients are lost when the bran and aleurone layer, which contains proteins, vitamins, and minerals, are separated. To compensate, wheat flour is often fortified to replace these lost micronutrients. The increase in surface area caused by size reduction may also lead to greater oxidation and a loss of nutrients, such as vitamin A and essential fatty acids.

Additionally, enzymes might be activated, speeding up spoilage and reducing shelf life. Using dull knives during size reduction can damage cell structures, leading to excess juice drainage and nutrient loss in fruits and vegetables. Size reduction does not inherently improve food safety and may actually increase contamination risks by exposing more surface area. Contaminated outer surfaces like skins, husks, and bran can transfer pathogens to the interior of the food. As with any process, size reduction on unsanitized surfaces can heighten food safety concerns.

QUALITY STANDARDS FOR VEGETABLES PRIMARY PROCESSING

- Different processing methods are used to remove inedible parts based on the edible parts of various raw materials.
- Vegetables should be washed using suitable methods to remove soil, insect eggs, pesticides, and other contaminants to ensure hygiene.
- Washing vegetables before cutting is essential to prevent nutrient loss and contamination.
- After washing, vegetables should be placed in a container that allows for drainage and is neatly arranged to facilitate cutting and further processing.

Primary Methods of Processing Vegetables

Leafy Vegetables

Leafy vegetables are those that use leaves and tender petioles for cooking. They can be categorized into three types ordinary leafy vegetables, bulbous leafy vegetables, and spice leafy vegetables. Common examples include spinach, rapeseed, and small day vegetables for ordinary leafy types Chinese cabbage and cabbage for headed leafy types; and leeks, celery, fennel, and coriander for spice leafy types.

- **Processing Method:** Remove any old or yellow leaves and wash the vegetables thoroughly with clean water. If insect eggs are present, soak the leaves in a light saltwater solution and then wash them again.

Stem Vegetables

Stem vegetables are those that have thick, tender stems or modified stems for cooking, such as lettuce, kale, bamboo shoots, garlic, shallots, water chestnuts, taro, lotus root, and ginger.

- **Processing Method:** Peel off the outer skin and rinse the vegetables with water to clean them.

Root Vegetables

Root vegetables are those that use enlarged roots as edible parts, like white radish, carrots, and yams.

- **Processing Method:** Scrub off any dirt from the surface, peel the outer skin with a peeler, and wash with clean water. After washing, soak the vegetables in water to prevent discoloration.

Fruits and Vegetables

Fruits and vegetables use fruits and young seeds for cooking, including eggplant, tomato, pepper, cucumber, zucchini, wax gourd, tender broad beans, and edamame.

- **Processing Method:** Peel the skin using a peeler and wash the vegetables. For thin-skinned varieties, briefly scald them in boiling water before peeling. When processing beans, remove the tough tendons from both sides and wash thoroughly.

Cauliflower

Cauliflower and broccoli are the common vegetables that use flowers as edible parts.

- **Processing Method:** Remove the outer leaves, trim off any discolored parts with a knife, separate the cauliflower florets, and wash them thoroughly with clean water.

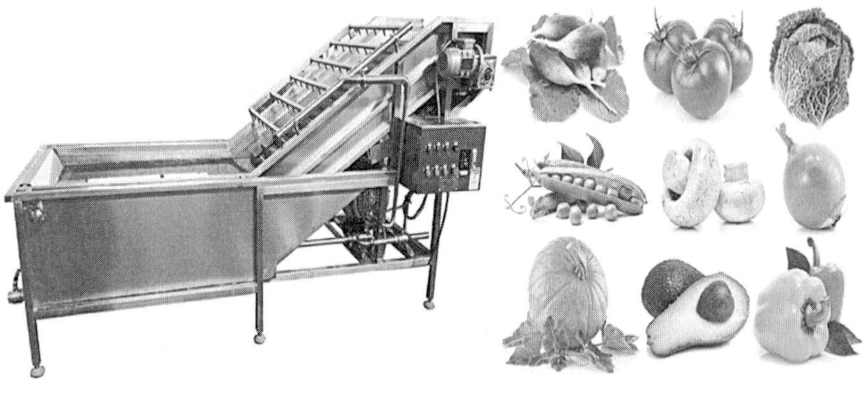

Figure: 1.3 Primary Processing of Vegetables

The methods described above are suitable for primary processing of vegetables with small-scale output. However, for factory-based vegetable processing, it's more efficient to use a vegetable processing line. These lines can handle a variety of vegetables and typically require only the replacement of a specific vegetable washing machine for different types of produce. As a manufacturer and supplier of vegetable processing lines with years of experience, we have exported to over 180 countries worldwide.

Fruit and Vegetable Processing

Step 1. Inspect the Fruits and Vegetables that Arrive

Workers should inspect fruits and vegetables for diseases, mold, rotting, and other issues before they leave the farm. By removing damaged crops, healthy ones are protected, though there's always a chance something may be overlooked. To ensure quality, factory staff should conduct a secondary inspection when the new crops arrive, carefully checking each item for defects to prevent damaged produce from contaminating the rest.

Step 2. Sanitize your Equipment Regularly

Other equipment in a processing plant, however, does not have antibacterial properties. Bacteria and other harmful pathogens are more likely to cling to these surfaces, potentially spreading to the fruits and vegetables being processed. Regular cleaning and sanitizing of this equipment are essential to prevent bacterial contamination. Even with the antibacterial qualities of certain metals, properly cleaning stainless steel equipment is crucial for maintaining a hygienic and safe processing environment.

Step 3. Employees Should be Trained and Educated

Proper training is crucial to ensure smooth operations in fruit and vegetable processing factories. All personnel must undergo comprehensive training and education on their specific roles within the facility before being allowed on the production floor. In the event of staff shortages, employees should be capable of handling tasks beyond their primary responsibilities.

Training and education should go beyond just the physical tasks in the facility. Employees need comprehensive training on safety and sanitation protocols to ensure they fully understand how to maintain a safe and clean workplace.

Step 4. Use Hygienic Procedures

Hygiene plays a crucial role in maintaining sanitation in fruit and vegetable processing plants. To prevent the spread of bacteria and germs—particularly those introduced from outside the facility—employees must adhere to strict hygiene practices within the plant. Key hygiene practices include regular hand washing, wearing gloves and shoe covers, and keeping the facility clean. Ensuring both personal and facility-wide hygiene standards are met should be a top priority, so staff understands the expectations and can follow hygiene protocols effectively.

Step 5. On the Factory Floor, use a Face Mask and Hairnets

Employees should follow various hygiene precautions, such as wearing gloves and shoe covers. Face masks and hairnets are also essential for maintaining good sanitation within the facility. To help prevent the spread of germs and contamination, all personnel should be required to wear face masks and hairnets. As an extra safety measure, men with facial hair should wear hairnets to cover their beards.

Step 6. Prepare a Pest-control Strategy

Pests, including insects and rodents, are a common issue in food and beverage establishments. These pests can carry diseases and contaminants, posing significant health risks. It's essential to have a pest management plan to prevent infestations. Keeping the facility clean and making sure no food residues are left out where pests could be attracted are key steps in pest prevention. Employees should also ensure that fruits and vegetables are stored properly to avoid attracting pests. Pest control also involves sealing any gaps that could allow pests to enter the facility and installing deterrents on the exterior.

Primary processing of fruits and vegetables is a key step in delivering high-quality, nutritious produce. From cleaning and sorting to packaging and preservation, these processes are essential for maintaining the freshness and nutritional value of the fruits and vegetables we enjoy. So, the next time you bite into a crisp apple or enjoy a fresh salad, take a moment to appreciate the detailed journey these items took through primary processing to reach your plate – a journey that not only boosts their flavor but also their health benefits.

2

Operation and Maintenance of Washer and Graders

Washing is typically the first step in the fruit and vegetable production process. In small-scale operations, this is usually done in basins with recirculating water or with still water that is regularly replaced. The goal of washing is to remove dirt and contaminants from the produce before it enters the processing line, preventing potential contamination of the raw materials.

The washing should be done with clean water, as pure as possible, and if needed, the water can be made potable by adding 10 ml of a 10% sodium hypochlorite solution for every 100 liters of water. It's also important to use tools that ensure thorough cleaning of the produce, leaving no dirt behind for the next stages of processing.

THE VEGETABLE WASHING MACHINE'S PRINCIPLE

A vegetable washing machine was designed and manufactured with the needs of micro, small, and medium-scale vegetable growers in mind. The machine focuses on high washing capacity and efficiency, with stainless steel components used to ensure the quality of the washed produce. A durable primary frame was incorporated to maintain the machine's structural integrity.

To enhance safety and ease of use, protective coverings were added around the moving parts, while also prioritizing ergonomic design and straightforward operation and maintenance. The machine features rollers for easy mobility, and as the vegetables move along the conveyor belt at a set speed, sprinklers spray water to clean them effectively.

This system ensures that the vegetables are thoroughly cleaned to the required quality. High-pressure sprinklers spray water across and along the conveyor belt, helping to clean the vegetables. During operation, the vegetables remain stationary on the conveyor belt, with no relative motion between the produce and the belt, minimizing any potential damage until the cleaned produce reaches the dryer sheet, where it is released into a collector. The waste and unwanted particles are removed using a drain net that is dissolved in the washing water.

To conserve water, the water used in the washing process is recycled. After the cleaning, it is pumped and filtered back into the washing chamber. The vegetable washing machine incorporates a spray cleaning function during the material transfer process, which helps replace the dirty water on the produce's surface, enhancing the cleaning effect. After secondary cleaning with the external spray water, the water automatically flows into the water tank, replacing the cleaning tank and maintaining water quality for continued use. This improves cleaning efficiency and enhances the overall cleaning result.

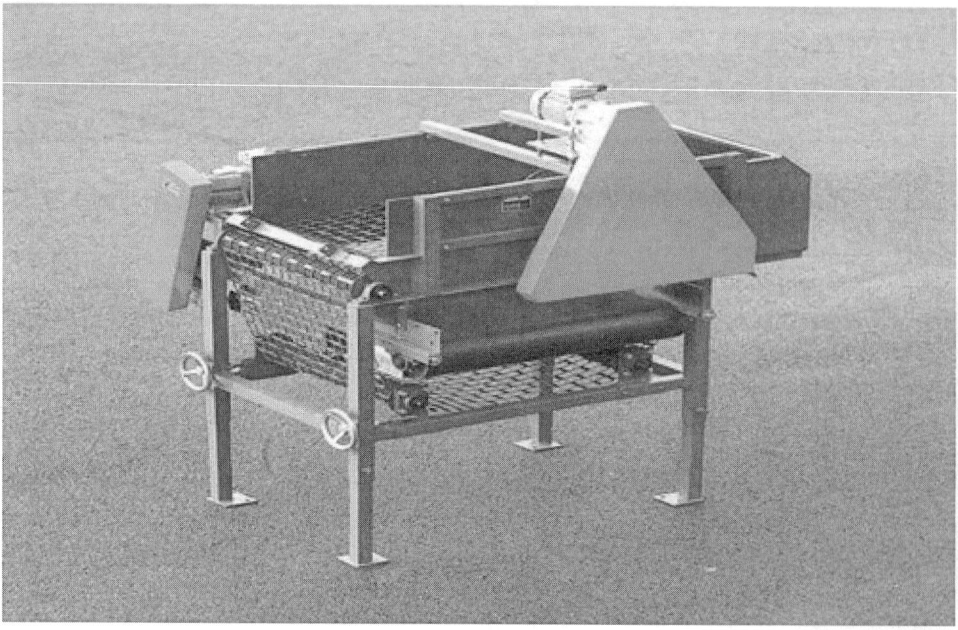

Figure: 2.1 Vegetable washing machine

Additionally, a special feature in the leaf vegetable bubble washer allows the vegetables to have more contact with water and air, while the pushing mechanism moves lightweight leafy vegetables at a consistent speed. Water is pumped by the circulating water pump, allowing contaminants to enter the water circulation tank, where precipitated impurities are treated. The filtration device in the water tank can be easily dismantled, leaving the filtered contaminants in the tank. Floating debris is discharged through the upper overflow port, ensuring timely removal of contaminants and maintaining the cleanliness of the water.

ADVANTAGES OF VEGETABLE WASHING MACHINE

The vegetable washing machine is constructed with a support frame and conveyor belt made from 304 stainless steel, ensuring stability, ease of maintenance, and a long service life. The entire machine's running speed is adjustable, and the conveyor speed can be set according to production requirements, offering high automation, labor savings, and enhanced production efficiency. With the dual action of air bubbles and spray, the machine provides multidirectional cleaning at various angles. The conveyor belt motor allows for variable speed adjustment.

Once the vegetables have been cleaned, they are automatically released after passing through the network chain. The conveyor belt's incline can be adjusted based on production needs, contributing to excellent operational efficiency. This system mimics manual cleaning actions, providing effective protection for the produce, avoiding bumps and scratches during the cleaning process, and offering water-saving benefits. The water circulation and filtration system can reduce water usage by up to 80%, making it eco-friendly.

Figure: 2.2 The vegetable washing machine's support from is made of stainless steel.

The machine also features a large sewage valve for efficient discharge and easy cleaning and maintenance. Quick-connect fittings make disassembly and cleaning hassle-free. The washer is portable, electric-powered, and constructed from durable stainless steel. The small gap between the conveyor components and the sink simplifies cleaning. The inner rotational drum is made of 1.5 mm thick stainless steel, measuring 760 mm in length and 620 mm in diameter. It is connected to two bearings via a hollow shaft, with a stainless steel water pipe installed inside. A water injection pump sprays pressurized water through the perforated inner shaft, ensuring thorough washing. The machine is equipped with a timer and an electrical system that controls the drum's rotational speed, which can be accurately adjusted up to 60 rpm. It also includes systems for supplying water, as well as removing dirty water and sediment. Protective guards are installed around the rotating parts and moving belts for safety.

Necessity of Fruit Washer Maintenance

As a commonly used fruit processing machine, it can save time and energy, but regular maintenance is essential to maximize its efficiency. Consistent upkeep also helps extend the lifespan of fruit processing equipment. The question is: How should you properly maintain a fruit washer? Below are some tips to help you maintain and adjust fruit washing machines effectively.

Figure: 2.3 Fruit washer

Chain Adjustment

After the electric motor is turned off, you can check the chain's tightness by pressing the middle of the chain with your fingers. Normally, the elastic range should be between 4 mm and 9 mm. If the deformation exceeds this range, you will need to adjust the idle gear to tighten the chain.

Transmission Maintenance

To inspect, refuel, or replace the transmission gearboxes, first, turn off the drive motor and remove the inspection plug. If the oil level is low, add oil as needed. Be sure to perform all these steps in a no-smoking environment for safety.

Belt Adjustment

Once the electric motor has stopped, press the belt with your fingers. If the compression range falls between 7 mm and 12 mm, the belt is in a normal state. Otherwise, adjust the idle gear to achieve the standard tightness.

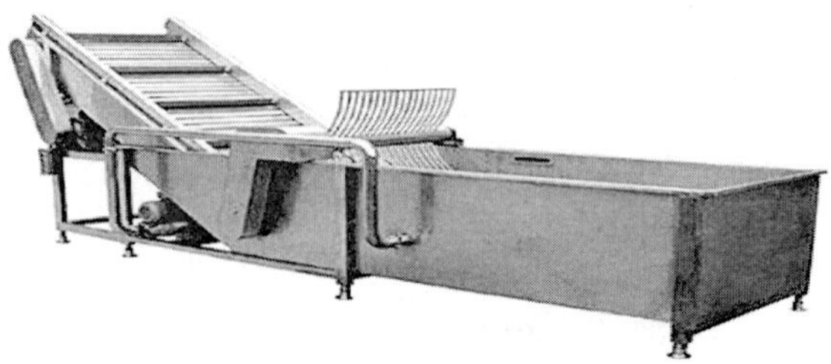

Figure: 2.4 Adjustment of belt

Preservation of Long-Term Unused Fruit Washer

If you're planning to store a fruit washer for an extended period, it's a good idea to clean off any dirt and wipe down the machine. Remove any debris from the belts and chains, and lubricate the rotating components and friction parts to prevent rust.

The Function of Ozone in Fruit Washing Machine

During the fruit washing process, the ozonator is activated and continuously generates ozone. The ozone gas is then delivered to the washing tanks through gas piping. When dissolved in water, the ozone creates ozone water, which plays a crucial role in sterilization, disinfection, and the breakdown of pesticides.

Figure: 2.5 Ozone in fruit washing machine

- **Degrade Pesticides, Hormones, and Toxins:** Ozone has strong oxidizing properties, allowing it to break down pesticide and hormone molecular chains instantly, converting them into stable, inorganic compounds. The oxygen atoms in ozone can also penetrate the cell walls of viruses and bacteria, helping to disinfect.
- **Separate Heavy Metal Ions:** The oxygen atom in ozone can oxidize dissolved heavy metal ions in water, turning them into insoluble, non-toxic compounds, effectively separating them.
- **Retain Freshness and Deodorize:** Ozone water or gas helps prolong the freshness of fruits and vegetables. Additionally, ozone gas can eliminate unpleasant odors in areas like bathrooms and kitchens.

GRADING OF FRUITS AND VEGETABLES

Agriculture products in particular makeup a significant portion of both domestic and foreign trades, sorting or classifying fruits and vegetables into various grades based on their size, shape, color and volume in order to sell them for a high price is known as grading.

Methods of Grading

Grading of fruits and vegetables is typically based on physical characteristics such as weight, size, color, shape, specific gravity, and freedom from diseases. For fresh market produce, grading is usually done manually or through machine grading. In both methods, size is the primary basis for grading. However, electronic grading systems are becoming more popular in the horticultural sector and have shown success in pilot studies. While the grading process is fully automated in some regions, it is still done manually in India.

The mechanical grading process usually involves a conveyor belt system with a collection bag at the end, allowing smaller produce to fall through, simplifying the grading process. Fruits and vegetables are generally graded according to local, federal, and international standards, with each country setting its own grading requirements based on market needs. For international markets, three main grades are typically used: Extra Class, Class I, and Class II.

- **Extra Class:** This grade represents superior quality, with produce exhibiting the correct shape and color for the variety, free from any internal defects that could affect texture and flavor. A tolerance of 5% is allowed for imperfections. The produce must be carefully arranged, with uniformity in size, color, and condition, and the packing or pre-packing materials should be of high quality and visually appealing.

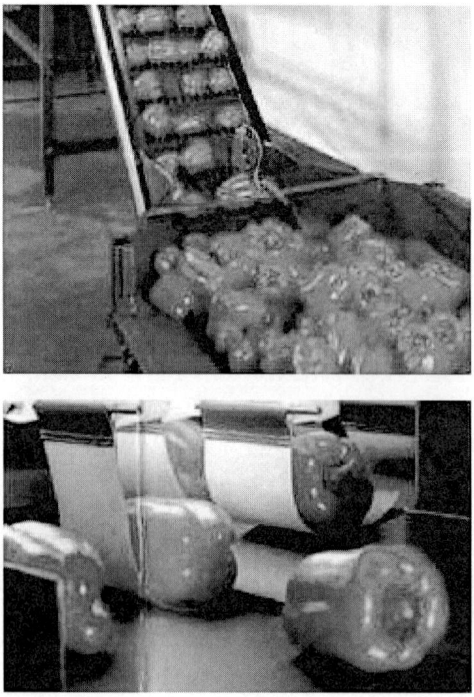

Figure: 2.6 Grading of capsicum

- **Class I:** This grade is almost identical in quality to Extra Class, but with a 10% tolerance allowed. Minor defects in shape, color, or skin are acceptable, as long as they do not impact the general appearance or the produce's keeping qualities.
- **Class II:** Produce in this class may have some external or internal defects, but it must still be fit for fresh consumption. Class II products are most suitable for local or short-distance markets, catering to customers who prioritize price over quality and are less concerned with imperfections.

Figure: 2.7 Different grades of kinnow

Fruits & Vegetables Grades as Recommended by the Directorate of Marketing and Inspection (DMI)

The Directorate of Marketing and Inspection (DMI), under the Ministry of Agriculture and Farmers Welfare, Government of India, was established in 1935 to develop grade standards in a scientific manner. The different grades used in some important fruits and vegetables with respect to weight, diameter or length as applicable are given as under.

Grapes

Grade	Size of berriers	
	Large bunch weight (g)	Small bunch weight (g)
Extra Class	200	150
Class I	150	100
Class II	100	75

Litchi

Grade	Diameter (mm)
Extra class	33
Class I	28
Class II	23

Kinnow

Size code	Diameter (mm)	No. of fruits in 10 kg pack
A	60-64	84
B	65-69	72
C	70-72	60
D	72-74	54
E	75-79	51
F	80-85	45

Apple

Size code	Weight (g)	Diameter (mm)
A	More than 241	More than 80
B	211-240	76-80
C	181-210	71-75
D	151-180	66-70
E	121-150	61-65
F	91-120	56-60

Guava

Size code	Weight (g)	Diameter (mm)
A	>350	>95
B	251-350	86-95
C	201-250	76-85
D	151-200	66-75
E	101-150	54-53
F	61-100	43-53

Mango

Grade	Weight (g)
A	200-350
B	351-500
C	551-800

Pear

Grade	Diameter (mm)
Extra class	60
Class I	55
Class II	50

Figure: 2.8 Grades used in Fruits

Potato

Size Code	Diameter (mm)
A	18.1-28.0
B	28.1-45.0
C	45.1-65.0
D	65.1-80.0
E	More than 80

Chillies

Size Code	length (cm)
A	less than 2.6
B	2.6 – 4.0
C	4.1 – 7.0
D	7.1 – 10.0
E	10.1 and above

Cabbage

Size Code	Weight (g)
A	201-600
B	601-1200
C	1201 and above

Tomato

Size Code	Diameter (mm)
1	30-34
2	35-39
3	40-46
4	47-56
5	57-66
6	67-81
7	82-102
8	102 and above

Onion

Size Code	Diameter (mm)
A	10-20
B	21-40
C	41-70
D	71 and above

Okra

Size Code	length (mm)
A	40.1 – 65.0
B	65.1 – 90.0
C	90.1 – 115.0
D	115.1 and above

Sweet pepper

Size Code	Diameter (mm)
Elongated (pointed)	20
Square (blunt) Square tapering (pegtop)	40
Flat (tomato pepper)	55

Cucumber

Size Code	Weight (g)
A	Upto 180
B	181-250
C	251 and above

Figure: 2.9 Grades used in vegetables

Figure: 2.10 Graded fruits and vegetable in market

Advantages of Grading

- Graded produce tends to achieve a higher price in the market.
- Grading builds stronger trust and confidence between buyers and growers.
- It enhances marketing efficiency by allowing produce to be bought and sold without the need for personal selection.
- Grading can help reduce high marketing costs related to packing and transportation.
- It boosts profits for distributors.
- It also increases profits for producers.
- Grading ensures product uniformity within each grade, which serves as the foundation for pricing.

3

Study of Refrigeration System and Freezing Equipment

Freezing is among the oldest and most commonly used techniques for preserving food, as it effectively maintains taste, texture, and nutritional value better than other preservation methods. This process leverages the advantages of low temperatures, which inhibit microbial growth, slow down chemical reactions, and delay cellular metabolic activities.

ROLE OF REFRIGERATION SYSTEMS FOR FRUITS AND VEGETABLES

Proper collection, handling, storage, and distribution of fruits and vegetables are crucial to ensuring they reach consumers in the best possible condition. Refrigeration plays a key role in maintaining freshness and extending the shelf life of these products. Companies involved in production, processing, and marketing that implement efficient commercial and industrial refrigeration systems benefit in several ways:

- Slowing down the ripening and aging of tissue cells.
- Preventing microbial growth.
- Reducing overall quantitative and qualitative losses.

Refrigeration systems for fruits and vegetables typically include cold storage rooms with controlled atmospheres. Environmentally friendly systems often use CO_2, glycol, or ammonia as refrigerants. These chambers regulate temperature, humidity, and air circulation to preserve product quality.

This process is particularly critical for climacteric fruits such as tomatoes, pears, apples, quinces, and melons. Before harvesting, factors such as collection timing, transportation, and storage conditions are carefully planned. These fruits are usually harvested before full ripening, and their maturation is later completed in cold storage facilities that maintain optimal conditions until they are ready for sale.

Variables to Control

In contrast, non-climacteric fruits, such as citrus varieties (oranges, lemons, and tangerines), as well as cucumbers and peppers, require a different approach. For these fruits, all handling, storage, and distribution processes are carefully planned, ensuring that they remain on the plant until they reach full maturity. Consequently, the key factors to regulate and monitor in the refrigeration of fruits and vegetables include temperature, relative humidity, and ventilation to maintain their quality and freshness.

The Cold Room's Ideal Tempenature

The temperature factor significantly affects the condition of fruits, depending on whether they are climacteric or non-climacteric. For climacteric fruits, proper temperature control helps slow down ripening, while for non-climacteric fruits, it prevents deterioration and extends shelf life.

Excessively high temperatures in cold storage can damage fruit tissue and disrupt enzymatic activity. If temperatures exceed 40°C or 60°C, fermentation may occur, leading to alcoholic flavors and changes in texture. Conversely, extremely low temperatures near or at the freezing point can cause unwanted alterations in taste and tissue integrity.

For tropical fruits, especially those sensitive to cold, exposure to temperatures between 5°C and 14°C for an extended period can result in blackening and flavor changes.

Due to these risks, precise temperature control in fruit refrigeration systems is essential. This is achieved using thermal probes with damping mechanisms, which help maintain stable internal conditions. In some cases, a control probe may be placed inside the fruit to monitor its internal temperature and track its condition over time.

Relative Humidity in Fruits and Vegetables Refrigeration Systems

Along with temperature, humidity is another crucial factor in the refrigeration of fruits and vegetables. During ripening, fruits naturally release moisture and lose water, leading to dehydration and faster aging. To prevent this, cold storage rooms must maintain a high relative humidity between 85% and 95%.

Ideally, the humidity levels should be 95% for vegetables and 90% for fruits. However, this varies depending on the surface-to-volume ratio of each product. Proper humidity control is essential, as improper regulation can lead to the growth and spread of mold.

Managing relative humidity is possible through control systems that regulate refrigerants like ammonia, CO_2, or glycol inside the evaporator. Additionally, moisture levels can be maintained by spraying water or using steam within the storage area.

Ventilation and Atmosphere for Product Conservation

Effective ventilation control is a critical aspect of fruit refrigeration systems. Proper ventilation ensures that the oxygen level does not drop below 2%, preventing an environment that could interfere with the ripening process. If oxygen levels become too low, fruits that require it for color development during ripening may remain green, leading to fermentation and potential spoilage.

To prevent this, cold storage rooms must regulate the balance of oxygen and CO_2 according to the specific needs of each fruit. Any variations in these gas levels are typically adjusted using nitrogen, which has no negative impact on the produce. Additionally, it is crucial to manage the ethylene gas naturally released by fruits during ripening. Proper ventilation systems help remove excess ethylene, ensuring controlled ripening and extending the shelf life of stored fruits.

Significance of Refrigeration Systems in Preserving Fruits and Vegetables During Distribution

Fruits and vegetables fall under the category of perishable goods, requiring specialized distribution methods supported by a well-organized logistics system. Maintaining a consistent temperature throughout their journey is essential to ensure that the cold chain remains unbroken and the products stay fresh.

Given these requirements, the transport and storage of fruits and vegetables must involve vehicles and facilities that meet specific operational criteria, such as:

- Capacity, weight, and dimension limitations.
- Fuel efficiency optimization.
- Structural durability and strength.
- Compatibility between the type of goods and the vehicle.
- Integration with other transportation modes like ships, trucks, or trains.
- Compliance with safety regulations for secure handling.

These transport unit characteristics also impact key factors such as:

- The cargo space volume and its organization.
- Engine power and efficiency.
- Layout and protection of storage areas.
- Crew comfort and operational ease.
- Ensuring product safety during transit.

To maintain cold chain integrity, the CCQI standard (Cool Chain Quality Indicator) is followed, which ensures professional temperature monitoring throughout the supply chain. This system is specifically designed for handling Perishable and Temperature-Sensitive Products (PTSP) by establishing quantitative quality indicators for individual operations and verifying compliance with cold chain regulations.

Storage Life of Frozen Foods

The International Institute of Refrigeration defines storage life as the period during which a food product can be stored under specific conditions and remain fit for consumption. The storage life varies across different food products, and it is important to note that it can be influenced by factors such as location and other environmental condition.

Table 3.1 Storage life of various food products

Food Product	Optimum Temperature (°C)	Relative Humidity (%)	Storage Life
Canned Goods	10-21	50-60	1-5 years
Pasta	15-21	50-60	1-2 years
Rice (white)	15-21	50-60	2-5 years

Food Product	Optimum Temperature (°C)	Relative Humidity (%)	Storage Life
Flour	15-21	50-60	6-12 months
Sugar	15-21	50-60	Indefinite
Honey	15-21	50-60	Indefinite
Dry Beans/Lentils	15-21	50-60	1-2 years
Oats	15-21	50-60	1-2 years
Crackers	15-21	50-60	6 months to 1 year
Cereals (boxed)	15-21	50-60	6-12 months
Potatoes (raw)	7-10	85-90	1-2 months
Onions (raw)	7-10	65-70	1-2 months
Garlic (raw)	7-10	65-70	3-6 months
Tomatoes (fresh)	18-22	85-90	1 week
Carrots (fresh)	0-4	90-95	2-3 weeks
Leafy Greens (fresh)	0-4	90-95	3-7 days
Cheese (hard)	4-8	70-80	3-6 months
Cheese (soft)	4-8	70-80	1-3 weeks
Milk	1-4	90-95	1 week
Eggs	1-4	80-85	3-5 weeks
Meat (beef, pork, lamb)	1-4	85-90	1-2 days (raw), 6-12 months (frozen)
Poultry (chicken, turkey)	1-4	85-90	1-2 days (raw), 12 months (frozen)
Fish (fresh)	-1-4	85-90	1-2 days (fresh), 3-6 months (frozen)
Bacon	1-4	85-90	1-2 weeks
Frozen Vegetables	-18	80-90	10-12 months
Frozen Fruits	-18	80-90	10-12 months
Frozen Meals	-18	80-90	3-12 months
Bread (store-bought)	18-22	50-60	1-2 weeks
Bread (homemade)	18-22	50-60	1-2 days
Baked Goods (cookies, muffins)	18-22	50-60	1 week

Refrigeration Processes Involved in Food Processing

The core principle behind food preservation through refrigeration is to lower and maintain the food's temperature to minimize harmful or undesirable changes, which could be microbiological, physiological, biochemical, or physical. This process helps in preserving the nutritional quality of the food. Various refrigeration methods are employed during food processing to extend shelf life, and these methods are applied in succession as outlined below:

- **Pre-cooling:** Rapid cooling is essential to slow down the metabolism of food products and extend their lifespan. The lower the temperature, the better the quality and longevity of the food. Different refrigeration methods are used depending on the food type, such as hydrocooling for small fruits, vacuum cooling for leafy vegetables, blast-air systems for meat surfaces, specialized tanks for cooling milk and dairy products, and pulsed air for other products.
- **Chilling:** After pre-cooling, the food needs to be chilled and kept at an appropriate temperature. Chilling is typically carried out in cold rooms with proper ventilation. The temperature for chilling depends on the product's sensitivity. For highly sensitive items like mangoes, ginger, and sweet potatoes, temperatures < 8-12°C should be avoided to prevent metabolic disruptions. For medium-sensitivity items like tangerines, green beans, and potatoes, temperatures around < 4-6°C are suitable. For low-sensitivity products, a temperature range of 2-3°C, close to freezing, is ideal. In slaughterhouses, low temperatures achieved by blast chillers or cold storage rooms are necessary to prevent microorganism growth. Eggs can be stored at higher temperatures, as long as the shell remains intact, and dairy products require a temperature of 8-10°C to inhibit the growth of pathogenic microbes.
- **Freezing:** Freezing involves lowering the product's temperature below its freezing point, which halts the metabolism of fruits and vegetables. The longer the food stays frozen, the longer it can be stored.
- **Super-cooling and Super-chilling:** These are advanced techniques for food storage. Super-cooling involves lowering the temperature of food just below its freezing point without ice formation, while super-chilling refers to partial freezing of food at a temperature slightly below its freezing point.

In addition to the above techniques, other methods such as fat crystallization, cryoseparation of unwanted components, and cryoconcentration are also used for specific preservation needs.

The use of Refrigerants in Food Processing Industries

Refrigerants are the working fluids used in cooling systems. Refrigerant selection plays a vital role in any refrigeration industry. Therefore, it has been a real challenge to develop and implement refrigerants with desired thermodynamic properties and high energy and eco-efficient performance. The various refrigerent used in food processing sectors and their percentages are shown in Figure.

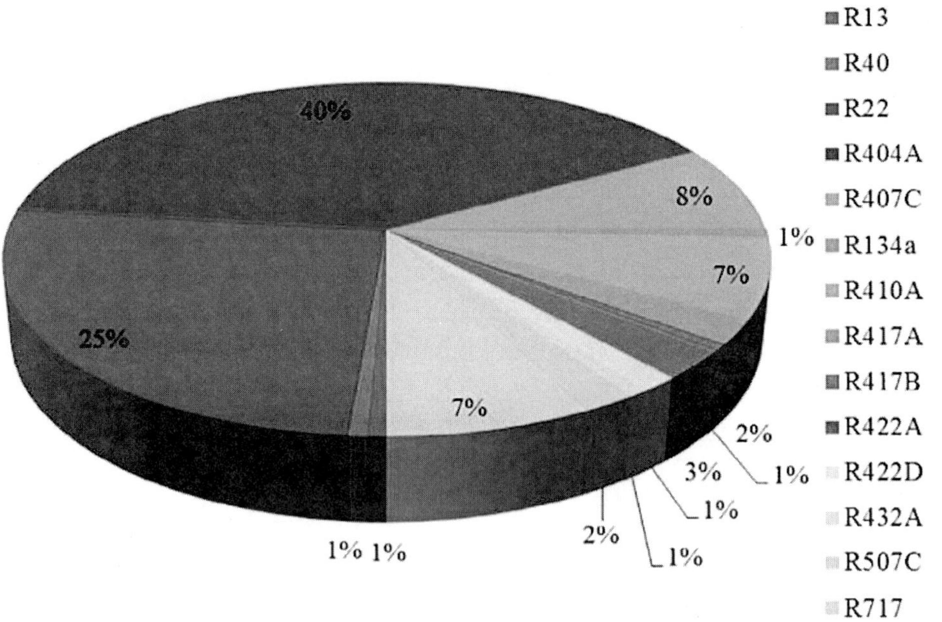

Figure: 3.1 Percentage distribution of refrigerants used in food processing

Previously, refrigerants were selected solely on the basis of their thermodynamic properties, but emphasis has also been placed on the ozone depletion potential (ODP) and global warming potential (GWP) of the refrigerants. Some of the desired properties of refrigerants required by the food processing industry are listed below:

- It should have low flammability and low toxicity.
- It should have the lowest ODP value and the highest GWP value possible.
- It should be easy to detect leaks.
- The choice of refrigerant has a significant impact on safety, reliability, energy consumption, system performance and also on the environment. Some of the refrigerants used in food processing industries are listed below in Table 3.2. However, some alternative refrigerants have now replaced the refrigerants used previously.

Table: 3.2 Table showing various refrigerants used in food indust

Type of industry	Earlier refrigerant used	GWP	Alternative refrigerant used	GWP
Small scale	R134a	1430	R744	1
Small scale	R22	1810	R1234y/R1234ze	4-6
Small scale	R404a	3920	Low GWP blends	<1300
Large Scale	R404a	3920	R717	0
Large Scale	R507a	3990	R744	1

In the past, chlorofluorocarbons (CFCs) and hydrochlorofluorocarbons (HCFCs) were widely used for their excellent thermodynamic properties. As an immediate substitute for CFC-12, R-134a was adopted due to its low ODP. However, there was a need for an alternative to R-134a as well, in order to address its high GWP. The composition of Table 3.3 alternative refrigerants are detailed in table.

Table: 3.3 Table showing compositions of 12 alternative refrigerants

Alternative Refrigerant	Composition (mass%)
1	R134aR1234yf/R32 (63/31/6)
2	R1234yf/R1234a (56/44)
3	R134a/R1234zeR1234yf (42/40/18)
4	R1234ze/R134a (58/42)
5	R1234yfR134a (60/40)
6	R1234ze/RR 134a (60/40)
7	R1234yf/R 152a/R134a (82/11/7)
8	R1234ze/R32/R152a (83/12/5)
9	R1234ze (100)
10	R1234yf (100)
11	R600a (100)
12	R600a/R290 (60/40)

Various types of refrigeration systems utilize different refrigerants based on their thermodynamic properties. It has been noted that R-404A is commonly used across many sectors due to its excellent performance in food conservation and freezing applications, as well as its widespread use in both split-type and centralized refrigeration systems.

General Recommendations on the Freezing Procedure

Freezing is a popular method of food preservation, offering numerous benefits in terms of maintaining food quality and simplicity of the process. Since the early days of freezing, the technology has evolved significantly due to advancements and improvements in freezing techniques. To fully understand and manage the concepts related to food freezing, it is essential to examine the fundamental factors that govern the freezing process.

Technology of Freezing

Freezing has long been a widely used preservation method, and its development has been shaped by technological advancements over time. The first significant milestone in freezing came in 1755, when a small quantity of ice was produced without using "natural cold." Initially, ice-salt systems were used to preserve fish, and by the late 1800s, freezing became a key method for commercial preservation. In the early stages, meat, fish, and butter were frozen in storage chambers and treated as bulk commodities.

Over the years, researchers and scientists made continuous efforts to improve commercial freezing techniques for various food products. Among these, fruits were especially significant, as freezing during peak harvest seasons helped preserve them for later use in products like jams, jellies, ice cream, pies, and other bakery items.

Today, freezing is the only large-scale method that bridges seasonal gaps and variations in the supply and demand of raw materials such as meat, fish, butter, fruits, and vegetables. It also enables the transportation of large quantities of food across vast geographical distances. To ensure the highest quality products, it is crucial to control the entire freezing process, from pre-freezing preparation to post-freezing storage. Understanding the theory of the freezing process and the parameters involved is essential for achieving the best results.

Process of Freezing

The freezing process primarily involves both thermodynamic and kinetic factors, which can influence each other at different stages of freezing. Significant thermal changes occur during freezing, marked by a reduction in the heat content of the material, as depicted in Figure. Initially, the material being frozen cools down to the temperature at which nucleation begins. Before ice can form, a nucleus, or seed, is required for the crystal to grow; this process of creating the seed is called nucleation.

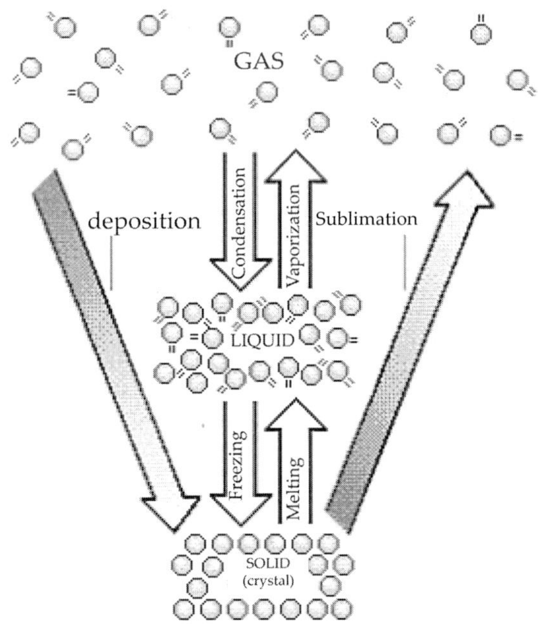

Figure: 3.2 A schematic illustration of overall freezing process.

Once the first crystal forms in the solution, a phase change occurs from liquid to solid, with further crystal growth. Therefore, nucleation is the initial step in the freezing process and is considered the critical phase that leads to a complete phase transition (Sahagian and Goff, 1996)..

Freezing Point of Foods

The freezing point is defined as the temperature at which the first ice crystal forms and the liquid is in equilibrium with the solid at that temperature. For pure water, this temperature is 0 °C (273 K). However, the freezing process of food systems is more complex due to the presence of both free and bound water. Bound water does not freeze, even at very low temperatures. Unfreezable water, which contains soluble solids, lowers the freezing point of water below 0 °C. As the freezing process progresses, the concentration of soluble solids increases in the unfrozen water, causing a shift in the freezing temperature. Consequently, the temperature at which the first ice crystal appears is often referred to as the initial freezing temperature. There are empirical equations available in literature that allow for the calculation of this initial freezing temperature for various foods based on their moisture content (Levy, 1979).

There are various methods of food freezing, and the quality of the frozen food can vary depending on the technique used. However, regardless of the method, the core principle behind all freezing processes remains the same in terms of the process parameters. The International Institute of Refrigeration (IIR) has established definitions to provide a framework for understanding the freezing process. According to their classification, the freezing process is divided into three stages based on significant temperature changes in specific areas of the product, as shown in figure for pure water and food, respectively.

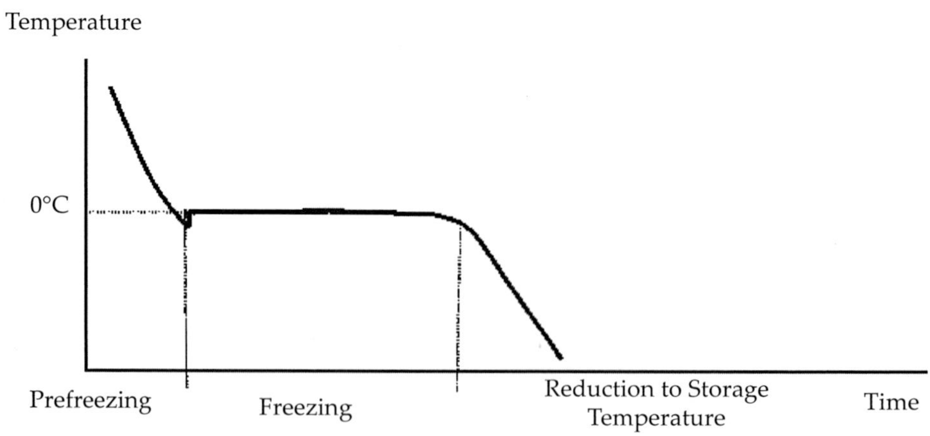

Figure: 3.3 Practical definition of the freezing process for pure water.

The first stage is the prefreezing stage, where the food undergoes the freezing process until the first ice crystal appears. In the case of pure water, the freezing temperature is 0 °C, and the temperature undergoes subcooling until ice formation begins. For foods, the temperature drops below the freezing point, and with the formation of the first ice crystal, it rises to the freezing point. The second stage, called the freezing period, involves the phase change from water to ice. For pure water, the temperature remains constant during this stage; however, for foods, it decreases slightly due to the increasing concentration of solutes in the unfrozen water. The final stage begins when the product temperature reaches the point where most of the freezable water has turned into ice, and it ends when the temperature is reduced to the storage temperature (Persson and Lohndal, 1993).

Freezing time and freezing rate are critical factors in designing freezing systems. The rate of freezing has a major impact on the quality of the frozen product, while freezing time is determined based on the freezing rate. These parameters are crucial for industrial applications when comparing various freezing systems and equipment (Persson and Lohndal, 1993).

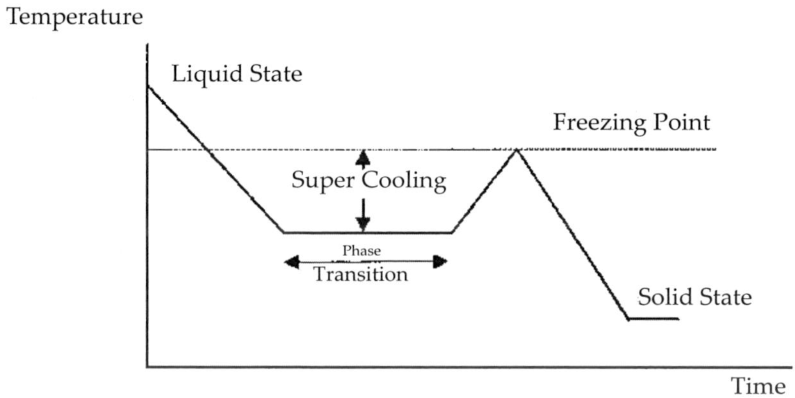

Figure: 3.4 Working definition of the food freezing process

Freezing Time

Freezing time is a crucial parameter in the freezing process, defined as the time required to lower the product's temperature from its initial value to a specific temperature at its thermal center. Since the temperature within the product varies during freezing, the thermal center is often used as the reference point. When the geometrical center of the product reaches the desired final temperature, it indicates that the average temperature of the product has been reduced to the storage temperature. Freezing time is influenced by various factors, such as the initial and final temperatures of the product, the amount of heat removed, the product's dimensions (especially its thickness), shape, heat transfer process, and the temperature of the refrigerating medium.

The International Institute of Refrigeration (1986) defines several factors that affect freezing time, both in terms of the frozen product and the freezing equipment (Persson and Lohndal, 1993). The most important factors include:

- Dimensions and shape of the product, particularly its thickness.
- Initial and final temperatures of the product.
- Temperature of the refrigerating medium.

- Surface heat transfer coefficient of the product.
- Change in enthalpy.
- Thermal conductivity of the product.

Calculating freezing time in food systems is more complex than in pure systems, as the freezing temperature varies throughout the process. A simplified approach treats the time elapsed from the start of freezing until the product is fully frozen as the freezing time. Plank's equation is commonly used to estimate freezing time, although due to the assumptions involved in the calculation, it provides only an approximation. The derivation of this equation assumes that the product being frozen starts at the freezing temperature, so the calculated time represents just the freezing period. The equation can be modified for different geometries such as slab, cylinder, and sphere, with coefficients adjusted according to the dimensions for each geometry (Plank, 1980).

Table: 3.4 Coefficients P and R of Equation

Geometry	P	R	Dimension
Infinite slab	1/2	1/8	thickness e
Infinite cylinder	1/4	1/16	radius r
Sphere	1/6	1/24	radius r

$$t_F = \frac{\rho\lambda_1}{T_F - T_e}\left[\frac{e^2R}{k} + \frac{eP}{h}\right]$$

where $l\gamma_1$ is the latent heat of frozen fraction, k and r are the thermal conductivity and density of the frozen layer, while h is the coefficient of heat transfer by convection to the exterior. T_f denotes the body temperature of the product when introduced into a freezer in wich the external temperature is T_e The coefficients R and P are given in Table 4 and arranged according to the geometry of the product frozen. where the letter e denotes the dimension (i.e. for infinite slab geometry, e is thickness of the slab and for infinite cylinder or sphere e is replaced by r which denotes the radius of the clylinder or sphere).

As mentioned earlier, Plank's equation assumes that the food is at freezing temperature at the beginning of the freezing process. However, the food is usually at a temperature above freezing. Therefore, the actual freezing time must be the sum of the time calculated by Plank's equation and the time required for the surface of the product to decrease from the initial temperature to the freezing temperature (Barbosa-Canovas and Ibarz, 2002). Several works have attempted to calculate the actual freezing time, such as that presented by Nagaoka et al. (1955). The Nagaoka equation calculates the amount of heat removed required to lower the temperature of a product from the initial temperature to the freezing temperature, as well as the amount of heat released during the phase change and the amount of heat removed to reach the freezing temperature. Other empirical equations can be found in detail in the literature.

$$t_F = \frac{\rho \Delta H}{T_F - T_e}\left[\frac{Re^2}{k} + \frac{Pl}{h}\right]\left[1 + 0.008(T_i - T_F)\right]$$

Where T_i is the temperature of the food at the start of freezing, ΔH is the difference between the enthalpy of the food at the initial and final temperatures of freezing. Re and Pl are dimensionless numbers, while k and h are the thermal conductivity and the heat transfer coefficient, respectively.

To calculate the freezing time of irregularly shaped products, a common property of most food products - especially fruits and vegetables - a dimensionless factor is used in the equation.

Rate of Freezing

The freezing rate (°C/h) of a product or package is defined as the ratio of the difference between the initial and final temperatures of the product to the freezing time. At any given location within the product, the local freezing rate can be described as the ratio of the difference between the initial temperature and the desired temperature to the time taken to reach the final temperature. The quality of frozen products is significantly influenced by the freezing rate.

Generally, rapid freezing results in higher quality frozen products compared to slow freezing. When freezing occurs rapidly, multiple locations within the food will start crystallizing almost simultaneously. In contrast, with slow freezing, ice crystals grow at a slower pace, resulting in fewer nucleation sites and the formation of larger ice crystals. Larger ice crystals can cause mechanical damage to cell walls and lead to cell dehydration. Therefore, the freezing rate is particularly crucial for plant tissues, as it affects ice crystal size, cell hydration, and damage to cell walls. The figure illustrates the typical behavior of the freezing preservation dynamics curve.

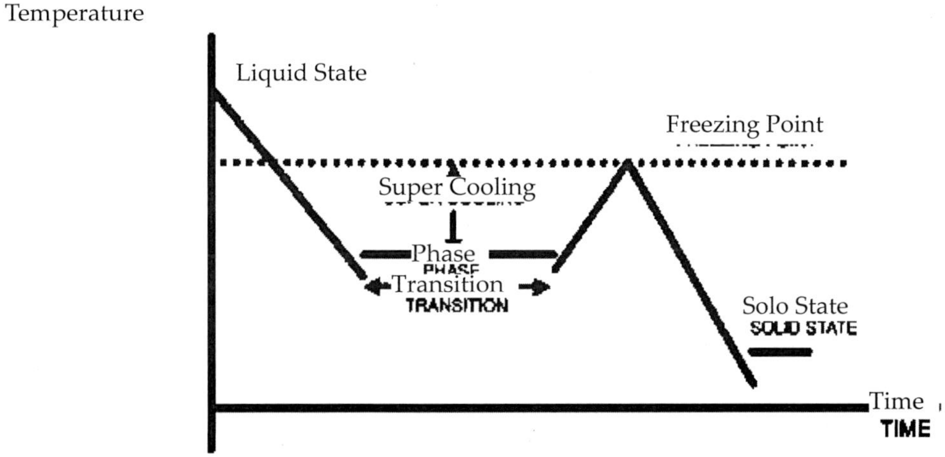

Figure: 3.5 Freezing preservation dynamics curve.

While rapid freezing is beneficial for most foods, some products may be prone to cracking when exposed to extremely low temperatures for extended periods. Various mechanisms, including volume expansion, contraction and expansion, and the buildup of internal pressure, have been proposed in the literature to explain the damage mechanisms during freezing (Hung and Kim, 1996).

Refrigeration

Refrigeration is the process of removing heat from a material that has a temperature higher than its surroundings. This mechanism is a key component of the freezing process and freezing storage, which involve the thermodynamic aspects of freezing. According to the second law of thermodynamics, heat naturally flows from areas of higher temperature to those of lower temperature. Consequently, to move heat from a lower temperature to a higher one, work must be performed. The goal of industrial refrigeration processes is to transfer heat from low-temperature points to areas with higher temperatures. To achieve this, the food industry commonly uses either closed mechanical refrigeration cycles, where refrigerants circulate, or open cryogenic systems using liquid nitrogen (LIN) or carbon dioxide (CO_2).

The key components of a closed mechanical refrigeration system include the condenser, compressor, evaporator, and expansion valve. Refrigerants such as hydrochlorofluorocarbon (HCFC) and ammonia are often used in these mechanical refrigeration systems. A simple diagram of the closed mechanical refrigeration system is shown in Figure.

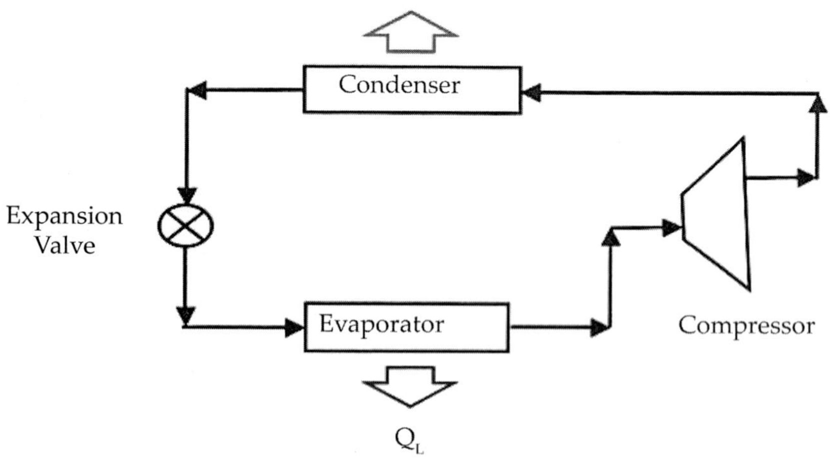

Figure: 3.6 A simple scheme for a one-stage closed mechanical refrigeration system.

The refrigeration process begins at the suction point of the compressor, where the fluid, initially in a vapor state, is compressed. This compression causes an increase in both the temperature and pressure of the fluid. The heated fluid then moves through the condenser, where it loses energy by releasing heat and changes into a liquid state. Once the phase change occurs inside the condenser, the fluid passes through the expansion valve, which causes a decrease in pressure, converting the liquid into a liquid-gas mixture. Finally, the mixture flows through the evaporator, where it transforms into a saturated vapor and absorbs heat from the surrounding environment as it cools. After this stage, the loop restarts.

Another refrigeration system used by the food industry is the cryogenic system, which employs carbon dioxide or liquid nitrogen. Unlike the circulating fluids in closed mechanical systems, the refrigerant in cryogenic systems is consumed in a different manner.

Refrigerants

There are various refrigerants available for refrigeration systems, and the selection of an appropriate refrigerant depends on the physical, thermodynamic, and chemical properties of the fluid. Environmental factors also play a crucial role in refrigerant selection, as leaks in the system can have harmful effects on the atmospheric ozone layer. Certain refrigerants, such as halocarbons, have been banned to prevent potential environmental hazards. For industrial purposes, ammonia is commonly used, while chlorofluoromethane and tetrafluoroethane are also recommended as refrigerants.

Freezing Capacity

The choice of freezing equipment is based on the need to freeze a certain amount of food per hour. For each type of freezer, the freezing capacity (expressed in tons per hour) is defined as the ratio between the amount of product that can be loaded into the freezer and the time that the product is held in that particular freezer.

The first parameter, the amount of food products loaded into the freezer, is influenced by the dimensions of the product and the mechanical limitations of the freezer. The denominator (holding time) plays an important role in freezing systems and is based on the calculation of the amount of heat removed from the product per hour, which varies depending on the type of product being frozen.

Freezing Systems

There is a range of freezing systems available for use, and for most products, multiple types of freezers can be employed. Therefore, when selecting a freezing system, it is essential to conduct a cost-benefit analysis considering three main factors: economics, functionality, and feasibility.

Financial considerations primarily involve the capital investment and the production cost of the chosen equipment. It is also important to account for product losses during the freezing process, as selecting more expensive freezers may reduce product losses, thereby offering long-term benefits.

Functional factors mainly focus on the compatibility of the selected freezer with the specific product. The freezing process mode, whether in-line or batch, should be taken into account, with computerized systems being more beneficial for ease of operation and lowering production costs. Additionally, mechanical constraints of the freezer should be considered, as some freezer types may not be physically suitable for certain products.

Lastly, feasibility involves evaluating factors like plant location, processing area location, cleanability, hygienic design, and the desired product quality.

These factors help narrow down the choices for selecting the appropriate freezer, but their relative importance may vary depending on the specific process. For developing countries, where freezing technology is still relatively new, cost becomes a primary factor due to the need for lower capital investment and the relatively lower production rates.

Freezing Equipment

Industrial freezing equipment can be classified in several ways, including by its operation type (batch or in-line), heat transfer systems (such as air, contact, or cryogenic), and its impact on product stability. The rate at which heat is transferred from the freezing medium to the product is a key factor in determining the freezing time. As a result, the choice of equipment for the freezing process directly influences the rate of freezing.

Air-Blast Freezers

The air blast freezer is one of the oldest and most commonly used freezing equipment due to its temperature stability and versatility for a wide range of products. Typically, air is used as the freezing medium in these systems, either as still air or forced air. Freezing is achieved by placing the food in freezing rooms known as sharp freezers. Still air freezing is the most cost-effective method, offering the added benefit of maintaining a constant temperature during frozen storage, which makes it suitable for unprocessed bulk products like beef quarters and fish. However, it is also the slowest freezing method due to the low surface heat transfer coefficient of circulating air inside the room. The freezing time in sharp freezers depends largely on the temperature of the freezing chamber, as well as the type, initial temperature, and size of the product.

An improved version of the still air freezer is the forced air freezer, where air is circulated through convection inside the freezing room. Even with modifications like additional refrigeration capacity and fans to increase air circulation, controlling the airflow over the products during slow freezing remains a challenge. A typical design for air blast freezers. is shown in Figure.

Figure: 3.7 Air blast freezer.

There are many designs and configurations for air blast freezers, primarily categorized by the mode of operation: inline or batch. Continuous freezers are best suited for large-scale production of packaged products that require similar freezing times, where the product is moved through the system on trucks or conveyors. When trucks are used, the system operates on a semi-batch principle, with trucks remaining stationary during the process until a new truck enters at one end of the tunnel, moving the others along to release a finished one at the exit. Batch freezers offer more flexibility, as they can freeze a variety of products at once on individual trolleys. However, overloading can be an issue with these freezers, requiring closer supervision compared to continuous systems.

Tunnel Freezers

In tunnel freezers, products on trays are placed in racks or trolleys and frozen through the circulation of cold air inside the tunnel. To facilitate proper air circulation, sufficient space is maintained between the layers of the trolleys, which can be moved continuously in and out of the freezer either manually or by forklift trucks. This freezing system is versatile and suitable for a wide range of products. However, it does come with some mechanical limitations, such as the need for high manpower for handling, cleaning, and transporting trays (Mallett, 1993). A typical trolley for a tunnel freezer.

Figure: 3.8 Trolley in a tunnel Freezer

Belt Freezers

Belt freezers were originally designed to ensure continuous product flow using a wire mesh conveyor inside the blast rooms. Early belt freezers faced challenges due to poor heat transfer mechanisms and mechanical issues, but modern versions have improved these problems by introducing vertical airflow, which forces air through the product layer. Effective airflow is achieved when the product is evenly distributed across the conveyor belt. To reduce floor space requirements, belts can be arranged in multi-tier or spiral belt freezers.

In a spiral belt freezer, the conveyor belt is designed to wrap around a rotating drum, maximizing the surface area of the belt in a limited floor space. This design eliminates the risk of product damage at transfer points, which is especially beneficial for products that require gentle handling. Spiral belt freezers are suitable for both packed and unpacked products with long freezing times (ranging from 10 minutes to 3 hours) due to the flexibility of the equipment.

Figure: 3.9 The cross-section view of a spiral belt freezer.

Fluidized Bed Freezers

The fluidized bed freezer, a type of blast freezer that has recently been modified for specific types of products, consists of a bed with a perforated bottom through which cold air is blown vertically upwards (Rahman, 1999). The system relies on cold air being forced from below the conveyor belt, causing the products to be suspended or floated in the cold air flow (George, 1993). The use of high air velocity is very effective for freezing unpackaged foods, especially when they can be completely surrounded by circulating air, as in the case of fluidized bed freezers. is shown in Figure.

Figure: 3.10 Cross-sectional view of a fluidized bed freezer.

The use of fluidization has several advantages over other freezing methods, as the product is individually quick frozen (IQF), which is suitable for particles that tend to stick together. The idea of individually quick frozen (IQF) foods arose with the first technological developments aimed at quick freezing. The need for an efficient way to freeze small particles that are likely to aggregate during the process is the purpose of IQF freezing. Small vegetables, shrimp, French fries, diced meat and fruits are some of the products currently frozen using this technology. A typical fluidized bed freezer is shown in Figure.

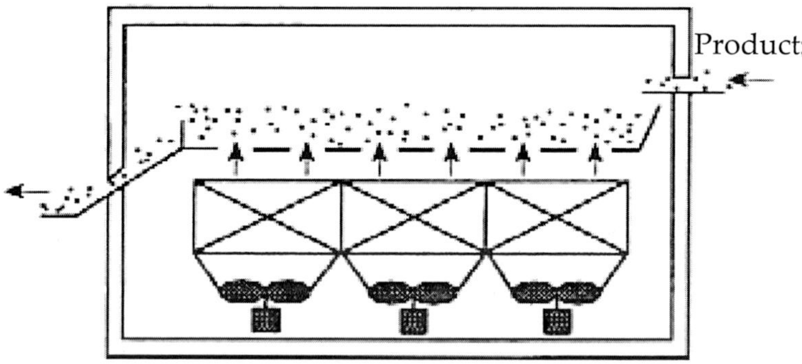

Figure: 3.11 Simple working principle of a fluidized bed freezer.

Contact Freezers

Contact freezing is one of the most efficient methods of freezing in terms of heat transfer. During this process, the product can either be in direct or indirect contact with the freezing medium. In direct contact freezers, the product is completely surrounded by the freezing medium, or refrigerant, which maximizes the heat transfer efficiency. A schematic illustration of this process is shown in Figure. In indirect contact freezers, the product does not come into direct contact with the freezing medium. Instead, it is in contact with a belt or plate that is, in turn, in contact with the freezing medium.

Figure: 3.12 Direct contact freezer

Immersion Freezers

The immersion freezer consists of a tank filled with a cooled freezing medium, such as glycol, glycerol, sodium chloride, calcium chloride, or mixtures of salt and sugar. The product is either immersed in this solution or sprayed while being conveyed through the freezer, leading to a rapid temperature reduction through direct heat exchange. Immersing a product into a liquid refrigerant is one of the quickest freezing methods, as liquids have better heat conductivity than air. The solute used in the freezing system should be safe, without any taste, odor, color, or flavor. For effective freezing, the product should have a higher density than the solution.

Immersion freezing systems are commonly used for shell freezing of larger particles, as they reduce the risk of product dehydration by quickly freezing the outer layer. However, a common issue with these systems is the dilution of the solution by the product, which can alter the concentration and process parameters. To prevent the product from directly contacting the liquid refrigerant, flexible membranes can be used.

Indirect Contact Freezers

In this type of freezer, the frozen materials are separated from the coolant by a conductive material, usually a steel plate. The mechanism of an indirect contact freezer is illustrated. Indirect contact freezers generally provide an efficient means of heat transfer, although the system has some limitations, particularly when used for packaged foods due to the resistance of the packaging to heat transfer. In addition, corrosive effects can occur due to the interaction of metal packaging with heat transfer surfaces.

Figure: 3.13 Indirect contact freezer

Plate Freezers

The most common type of contact freezer is the plate freezer. In this case, the product is pressed between hollow metal plates, either horizontal or vertical, with a coolant circulating through the plates.

Figure: 3.14 Plate freezer with a two-stage compressor and sea water condenser

This type of freezing system is typically used for regular-shaped materials or blocks, such as beef patties or block-shaped packaged products. A typical plate freezers in shown in Figure 3.14.

Contact Belt Freezers

This type of freezer is designed with a single or double belt for freezing thin layers of products. The design can be single or with a drum. Typical products frozen in belt freezers are fruit pulp, egg yolks, sauces and soups.

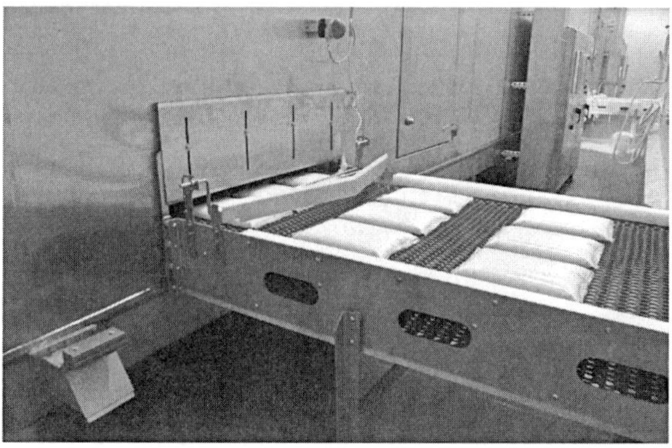

Figure: 3.15 Contact belt freezer

Cryogenic Freezers

Cryogenic freezing is a relatively modern freezing method in which food is exposed to an environment below -60 °C by direct contact with liquefied gases such as nitrogen or carbon dioxide (Hung and Kim, 1996). Unlike other freezing systems, cryogenic systems are not connected to a refrigeration plant. The refrigerants used are liquefied in large industrial facilities and then transported to the food-freezing plant in pressure vessels. The small size and mobility of cryogenic freezers provide design flexibility and improve the efficiency of the freezing process. While cryogenic freezers require low initial investment, they typically involve higher operating costs (Persson and Lohndal, 1993).

Liquid Nitrogen Freezers

Liquid nitrogen, with a boiling temperature of -196 °C at atmospheric pressure, is a by-product of oxygen production. This refrigerant is sprayed into the freezer, where it evaporates both as it exits the spray nozzles and upon contact with the products. The system is designed so that the refrigerant flows counter to the movement of the products on the belt, ensuring high heat transfer efficiency. The refrigerant consumption typically ranges from 1.2 kg of refrigerant per kg of product. Common food products processed using this system include fish fillets, seafood, fruits, and berries (Persson and Lohndal, 1993).

Liquid Carbon Dioxide Freezers

Liquid carbon dioxide can exist as either a solid or gas when stored at atmospheric pressure. When released at -70 °C, about half of the gas turns into dry-ice snow, while the other half remains in vapor form. This unique property of liquid carbon dioxide is utilized in various freezing systems, including a pre-freezing treatment that occurs before the product is exposed to nitrogen spray (George, 1993).

PACKAGING OF FROZEN FOOD

Proper packaging of frozen food is essential to protect the product from contamination and damage during transit from the manufacturer to the consumer, while also preserving its nutritional value, flavour, colour, and texture. Several factors are considered when designing suitable packaging for frozen food. The package should be visually appealing to the consumer, offer protection from external contamination, and be practical in terms of processing, handling, and cost (Rahman, 1999). The selection of packaging depends on the type and material used. Generally, there are three types of packaging for frozen foods: primary, secondary, and tertiary.

- Primary packaging is the packaging that comes in direct contact with the food, keeping the food enclosed until it is used.
- Secondary packaging is used to group primary packages together for sale.
- Tertiary packaging is used for bulk transportation of products (Harrison and Croucher, 1993).

The packaging material must be moisture-vapor-proof to prevent evaporation, thus maintaining the highest quality of the frozen food. Additionally, oxygen should be completely removed from the package using a vacuum or gas-flush system to prevent moisture and oxygen migration (ASHRAE, 1994; Sebranek, 1996). Glass and rigid plastic are examples of moisture-vapor-proof materials. While many packaging materials are not fully moisture-vapor-proof, some are sufficiently moisture-vapor-resistant to maintain acceptable food quality.

Common moisture-vapor-resistant packaging materials include most bags, wraps, and waxed cartons used in freezing packaging.

In general, containers should be leak-proof and easy to seal. Durability is another key factor, as packaging materials must not become brittle and crack at low temperatures.

A variety of packaging materials, generally categorized as rigid and non-rigid containers, can be used for primary packaging. Glass, plastic, tin, and heavily waxed cardboard are examples of rigid containers, often used for liquid food products. Glass containers are typically used for fruits and vegetables that are not water-packed. Plastics, derived from the oil-cracking industry, are also commonly used. Non-rigid containers include bags and sheets made of moisture-vapor-resistant materials like heavy aluminum foil, polyethylene, or laminated papers. Bags are the most commonly used packaging for frozen fruits and vegetables, due to their flexibility during processing and handling. These bags can be used alone or with outer cardboard cartons for added protection against tearing.

The shape and size of the container are also crucial in freezing products. Serving sizes may vary depending on the product type, and packaging should be selected based on the amount of food typically served in one meal. When considering the shape of the container, freezer space must be taken into account, as rigid containers with flat tops and bottoms stack efficiently, whereas round containers waste valuable freezer space.

Frozen Storage and Distribution

The quality of the final product depends on the history of the raw material. Using the lowest possible temperature is essential for the storage, transport and distribution of frozen products to obtain a high quality product, since the deterioration processes are mainly temperature dependent. The lower the temperature of the product, the slower the reaction rate, resulting in loss of quality. The temperature of the supply chain in refrigeration applications, from the factory to the retail store, must be carefully monitored.

FREEZING FRUITS AND VEGETABLES IN SMALL AND MEDIUM SCALE OPERATIONS AND ITS POTENTIAL APPLICATIONS IN WARM CLIMATES

Preservation of fruits and vegetables by freezing is one of the most important methods for maintaining the quality of agricultural products over a long storage period. In particular, the freshness of raw fruits and vegetables can be maintained for long periods, well beyond the normal growing season of most horticultural crops. Potential applications for freezing fruits and vegetables, including tropical products, have recently developed in parallel with developments in developing countries. The freezing of fruits and vegetables in small and medium-scale operations is detailed in the following sections and a general flow chart is shown in figure. 3.16.

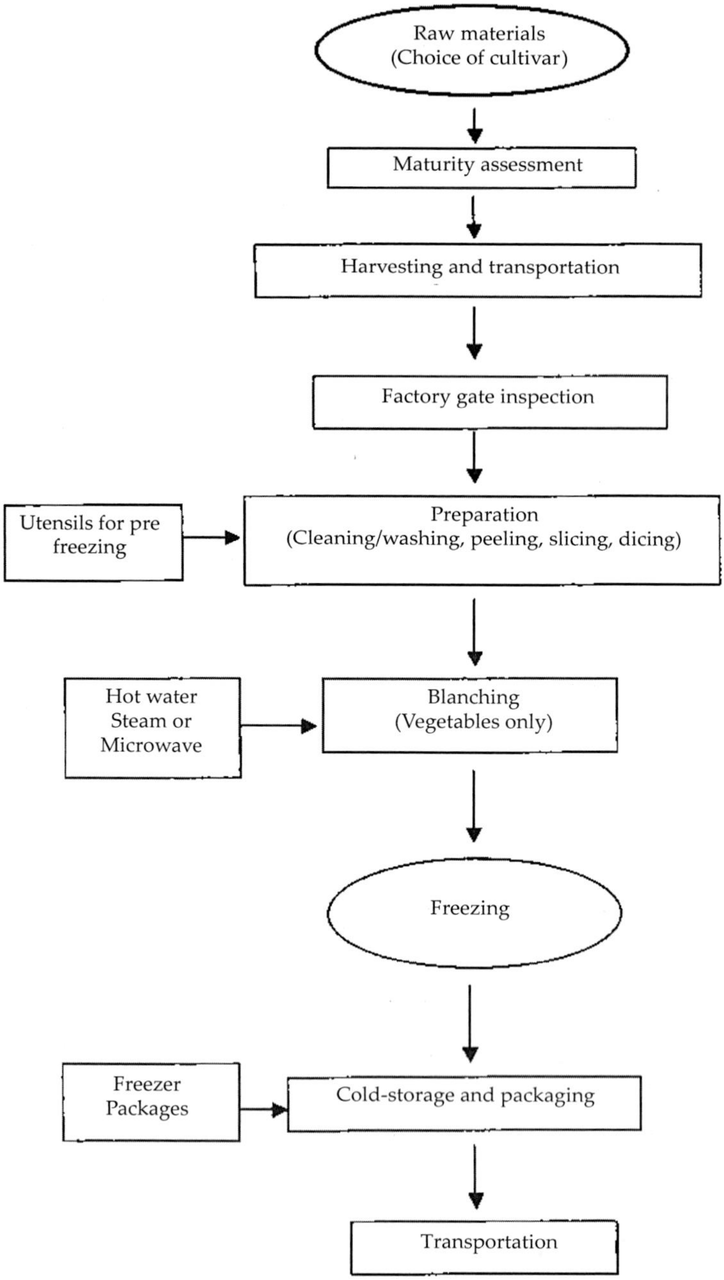

Figure: 3.16 A general flow chart of frozen fruits and vegetables

Freezing Fruits

The impact of freezing, frozen storage, and thawing on fruit quality has been studied for several decades. Today, frozen fruits are a significant and widely consumed food group (Skrede, 1996). The quality requirements for frozen fruit products largely depend on their intended use. If the fruit is meant to be eaten after thawing without further processing, texture characteristics become more important compared to when the fruit is used as a raw material in other industries. Conventional freezing methods typically damage the turgidity of living cells in fruit tissue. Unlike vegetables, fruits do not have a fibrous structure that can withstand this damage. Additionally, fruits are usually harvested in a fully ripe state, making them naturally soft in texture, whereas many vegetables are frozen while still in their immature state (Boyle et al., 1977).

Fruits also have delicate flavours, which can easily be altered or damaged by heat, making them ideal for raw consumption and less suitable for processing. Similarly, maintaining an attractive color is crucial for frozen fruits. To preserve quality, chemical treatments or additives are often used to inactivate the enzymes that cause deterioration. Therefore, proper processing is critical at every step, from harvesting to packaging and distribution. A freezing guide for freezing fruits is provided in Table. 3.5

Table: 3.5 Fruit freezing guide

Fruit	Preparation	Type of Pack
Apples	Wash, peel, and slice the fruit into an antidarkening solution made of 3 tablespoons of lemon juice per quart of water.	Pack in a 30-40% syrup, adding 1/2 teaspoon of crystalline ascorbic acid per quart of syrup. Alternatively, pack dry or with up to 1/2 cup of sugar per quart of apple slices.
Apricots	Wash, halve, and pit. Peel and slice if desired. If apricots are not peeled, heat in boiling water for 1/2 minute to keep skins from toughening during freezing. Cool in cold water, drain.	Pack in 40% syrup, adding 3/4 teaspoon crystalline ascorbic acid per quart of syrup.
Avocados	Peel soft, ripe avocados. Cut in half, remove pit, mash pulp.	Add 1/8 teaspoon crystalline ascorbic acid to each quart of puree. Package in recipe-size amounts.
Berries	Select firm, fully ripe berries. Sort, wash, and drain.	Use 30% syrup pack, dry unsweetened pack, dry sugar pack, (3/4 cup sugar per quart of berries), or tray pack.
Cherries (sour or sweet)	Select well-colored, tree-ripened cherries. Stem, sort, and wash thoroughly. Drain and pit.	Pack in 30-40% syrup. Add 1/2 teaspoon ascorbic acid per quart of syrup. For pies and other cooked products, pack in dry sugar using 3/4-cup sugar per quart of fruit.

Fruit	Preparation	Type of Pack
Citrus fruits, (sections or slices)	Select firm fruit, free of soft spots. Wash and peel.	Pack in 40% syrup or in fruit juice. Add 1/2 teaspoon ascorbic acid per quart of syrup or juice.
Grapes	Select firm, ripe grapes. Wash and remove stems. Leave seedless grapes whole. Cut grapes with seeds in half and remove seeds.	Pack in 20% syrup or pack without sugar. Use dry pack for halved grapes and tray pack for whole grapes.
Melons (cantaloupe, watermelon)	Select firm-fleshed, well-colored, ripe melons.Wash rinds well. Slice or cut into chunks.	Pack in 30% syrup or pack dry using no sugar. Pulp also may be crushed (except watermelon), adding 1 tablespoon sugar per quart. Freeze in recipe-size containers.

Production and Harvesting

The characteristics of raw materials play a crucial role in determining the quality of the frozen product. These characteristics include factors such as the genetic makeup, climate of the growing area, type of fertilization, and maturity of the harvest.

In terms of economic considerations for production, important fruit characteristics include ability to withstand rough handling, resistance to virus diseases and molds, uniformity in ripening, and yield. Mechanical harvesting often leads to bruising of the fruit and results in a broad range of maturity levels. In contrast, hand-picking allows for gentler handling and sorting by maturity, although it is generally more expensive due to high labor costs.

Harvesting fruits at their optimal maturity for commercial use can be challenging. Simple methods, such as pressure tests, are used to determine when fruits have reached their ideal maturity. Color is also an important indicator, as fruits darken as they mature. A combination of color and pressure tests is often considered the best method for assessing maturity levels.

Controlled atmosphere storage is commonly used for some fruits before freezing. This method involves storing fruits in an environment with high carbon dioxide and low oxygen levels, which slows the respiration rate, thereby extending the shelf life of respiring fruits. Since many fruits do not ripen significantly after picking, they are harvested as close to eating-ripe maturity as possible.

Pre-Process Handling and Operations

Freezing preservation of fruits can only maintain the inherent quality that the product originally had, as the freezing process does not enhance the quality characteristics of raw materials. Therefore, the quality level of the raw materials before freezing is the most important factor for successful freezing. Washing and cutting generally lead to losses when done after thawing, so fruits should be prepared (e.g., peeling, slicing, or cutting) before the freezing process. Freezing preservation does not require specific unit operations for cleaning, rinsing, sorting, peeling, and cutting fruits (Spiess, 1984).

Fruits that need to be peeled before consumption should undergo peeling prior to freezing. Peeling is typically done by scalding the fruit in hot water, steam, or hot lye solutions (Boyle and Wolford, 1968). The effect of peeling on the quality of frozen fruits has been studied for various types, including kiwi (Robertson, 1985), banana (Cano et al., 1990), and mango (Cano and Marín, 1992). The freezing rate can be improved by reducing the size of the fruits being frozen, especially for large fruits. A higher freezing rate leads to smaller ice crystals, which minimizes cellular damage in fruit tissue. Bananas, tomatoes, mangoes, and kiwis are examples of large fruits often cut into smaller cubes or slices before freezing (Skrede, 1996).

The main goal of blanching is to deactivate the enzymes that cause detrimental changes in color, odor, flavor, and nutritional value. However, the heat treatment involved in blanching can also cause the loss of some of these characteristics in fruits (Gutschmidt, 1968). As a result, only certain types of fruits are blanched to inactivate enzymes before freezing. Additionally, the loss of water-soluble minerals and vitamins during blanching should be minimized by maintaining the optimum combination of blanching time and temperature (Spiess, 1984).

Effect of Ingredients

The addition of sugars is a critical pretreatment for fruits before freezing, as it helps exclude oxygen from the fruit, which in turn preserves its color and appearance. When sugars are dissolved in solutions, they draw water out of the cells through osmosis, creating highly concentrated solutions inside the cells. The high concentration of solutes lowers the freezing point, reducing the amount of freezing that occurs within the cells and thereby preventing excessive structural damage. Typically, sugar syrups with a 30-60% sugar content are used to fully cover the fruit, serving as a barrier to oxygen transmission and browning. Various studies have demonstrated the protective effect of sugar on flavor, odor, color, and nutritive value during freezing, particularly for frozen berries.

Packaging

Fruits exposed to oxygen are prone to oxidative degradation, which leads to browning and a shortened storage life of the products. To prevent this, packaging of frozen fruits is designed to exclude air from the fruit tissue. Methods currently used to protect frozen fruits include replacing oxygen with sugar solution or inert gases, consuming oxygen with glucose-oxidase, and using vacuum and oxygen-impermeable films. Common packaging materials include plastic bags, plastic pots, paper bags, and cans, which are selected based on their penetration properties and thickness.

There are several types of fruit packs suitable for freezing: syrup pack, sugar pack, unsweetened pack, tray pack, and sugar replacement pack. The choice of packaging depends on the intended use of the fruit. Syrup-packed fruits are typically used for cooking, while dry-packed and tray-packed fruits are ideal for serving raw in salads and garnishes.

Syrup Pack

The proportion of sugar to water in a syrup pack depends on the sweetness of the fruit and the consumer's taste preferences. For most fruits, a 40% sugar syrup is recommended. Lighter syrups, which are lower in calories, are generally preferred for mild-flavored fruits to avoid overpowering their natural taste. In contrast, heavier syrups may be used for very sour fruits.

Syrup is made by dissolving sugar in warm water and then cooling the solution before use. The cooled syrup is added just enough to cover the prepared fruit, which is settled by gently shaking the container. To keep the fruit submerged in the syrup, a small piece of crumpled waxed paper or another water-resistant wrapping material is placed on top, pressing the fruit into the syrup before sealing the container and freezing it.

Pectin can be used to reduce the sugar content in syrups when freezing berries, cherries, and peaches. To prepare pectin syrup, dissolve 1 box of powdered pectin in 1 cup of water, stir, and boil for 1 minute. Then, add 1/2 cup of sugar, dissolve it, and cool the solution by adding cold water. The prepared fruit is placed in a 4 to 6-quart bowl, and enough pectin syrup is added to cover the fruit with a thin layer. The pack is sealed and promptly frozen.

Sugar Packs

In preparing a sugar pack, sugar is first sprinkled over the fruit, and the container is gently agitated until the juice is drawn out and the sugar dissolves. This type of packing is typically used for soft sliced fruits such as peaches, strawberries, plums, and cherries, with enough syrup added to cover the fruit. Some whole fruits may also be coated with sugar before freezing.

Unsweetened Packs

Unsweetened packs can be prepared in various ways, including dry-packed, covered with water containing ascorbic acid, or packed in unsweetened juice. When water or juice is used in syrup and sugar packs, the fruit is submerged using a small piece of crumpled water-resistant material. Generally, unsweetened packs result in a lower quality product compared to sugar packs. However, some fruits, such as raspberries, blueberries, scalded apples, gooseberries, currants, and cranberries, maintain good quality without the addition of sugar.

Tray Packs

Unsweetened packs are typically prepared using tray packs, where a single layer of prepared fruit is spread on shallow trays, frozen, and then promptly packaged in freezer bags. This method ensures that the fruit sections remain loose without clumping together, allowing for the advantage of using frozen fruit piece by piece.

Sugar Replacement Packs

Artificial sweeteners can be used as sugar substitutes to replace the sweet taste of sugar. While these sweeteners can mimic the sweetness of sugar, they cannot replicate the beneficial effects of sugar, such as color protection and the formation of thick syrup. Fruits frozen with sugar substitutes tend to freeze harder and thaw more slowly compared to fruits preserved with sugar.

Freezing Vegetables

Freezing is often regarded as the simplest and most natural method of preserving vegetables. Frozen vegetables and potatoes constitute a significant portion of the market for frozen food consumption. The quality of frozen vegetables is directly related to the quality of the fresh products, as freezing does not enhance the product quality. Pre-processing handling, from the moment vegetables are harvested to the time they are ready to eat, is a key factor in ensuring the retention of their quality.

COLD CHAIN REFRIGERATION

The term cold chain refers to a temperature-controlled supply chain through which a refrigerated product passes, remaining under specific temperature conditions until it is either retrieved by a customer in a retail setting or unloaded from a delivery vehicle at its destination. For consumers, the cold chain is typically associated with transport, retail, and household refrigerators.

However, refrigeration is also extensively used in the agri-food industry for storing raw materials, final products, and during food processing. An efficient and well-maintained cold chain ensures optimal conditions that prevent undesirable changes for as long as possible. Proper refrigeration ensures the safety and long shelf-life of food, maintaining its quality. A schematic diagram of a basic food chain is shown in figure.

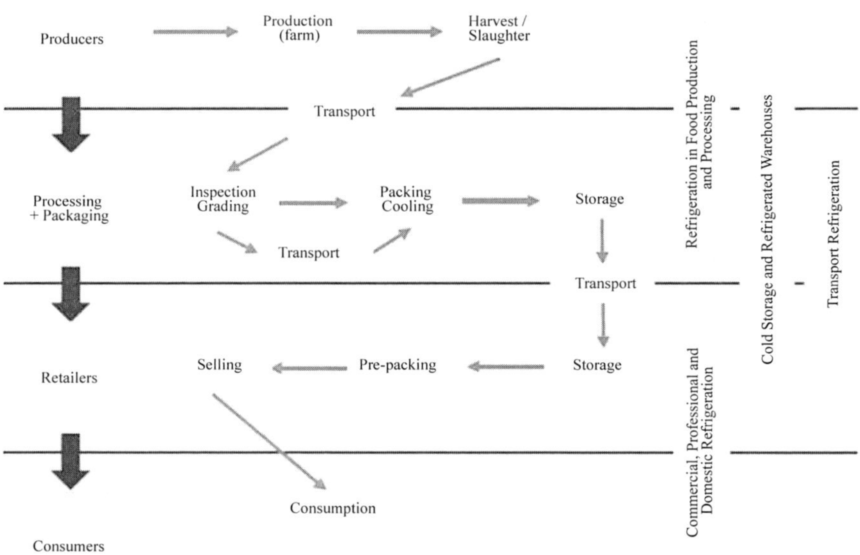

Figure: 3.17 Showing an overall cold chain and its components

As food progresses through the cold chain, managing and regulating its temperature becomes a challenging task. This is because the temperatures of bulk packs in large storage rooms are much less sensitive to small heat inputs compared to individual consumer packs in open display cases or in a domestic refrigerator.

Role of Refrigeration Cycle in Food Processing Industry

In food processing industries, refrigeration systems typically operate using a single-stage, direct expansion vapor compression refrigeration (VCR) cycle. These systems are simple, reliable, and cost-effective. A VCR cycle consists of four main components: compressor, evaporator, expansion valve, and condenser. In this system, a low-pressure cold liquid refrigerant evaporates into a gas within the evaporator coil, absorbing heat in the process, which cools the surrounding medium. The low-pressure hot gas from the evaporator is then compressed in the compressor to a high-pressure hot gas. This high-pressure hot gas passes through another coil, where it condenses back into a high-pressure cold liquid, releasing heat into the surrounding medium. The high-pressure cold liquid refrigerant then moves through the expansion valve, where the throttling process occurs, lowering its pressure. It is important to note that the throttling process in the expansion device is irreversible. The low-pressure liquid then returns to the evaporator. The schematic diagram of a basic mechanical VCR system and its corresponding T-s diagram have been shown in Figure. 3.18 Factors influencing the nutrient content of refrigerated food include storage temperature, storage duration, humidity, and light.

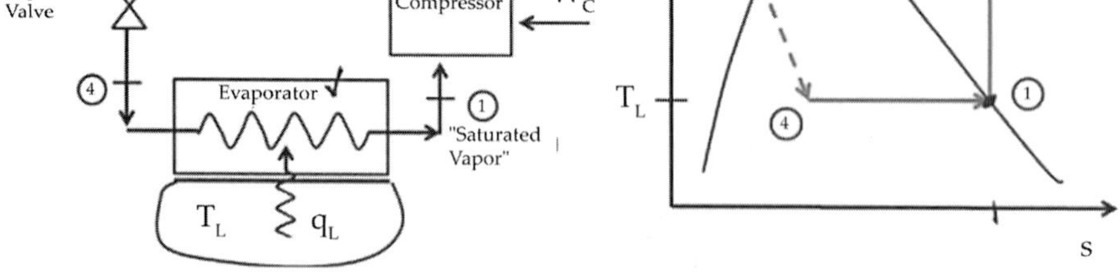

Figure: 3.18 (a) Schematic and (b) T-s diagram of a VCR system

The various processes involved in a VCR cycle with respect to the T-s diagram are as follows:

- Process 1-2: Isentropic work input to the compressor.
- Process 2-3: De-superheating and condensation at constant pressure.
- Process 3-4: Throttling in the expansion valve. Since the throttling process is irreversible, it is represented by a dotted line on the T-s diagram.
- Process 4-1: Evaporation at constant pressure.

However, these systems have specific requirements regarding the processing methods and component types used. For chilling, single-stage systems with flooded evaporators and screw compressors are recommended. Secondary refrigerants like propylene glycol are preferred over primary refrigerants, as they significantly reduce the quantity of ammonia and enhance safety. For freezing, two-stage refrigeration systems with flooded evaporators and either piston or screw compressors are preferred. Refrigerants such as R134a or R717 are used in the high-temperature stage, while R744 is employed in the low-temperature stage.

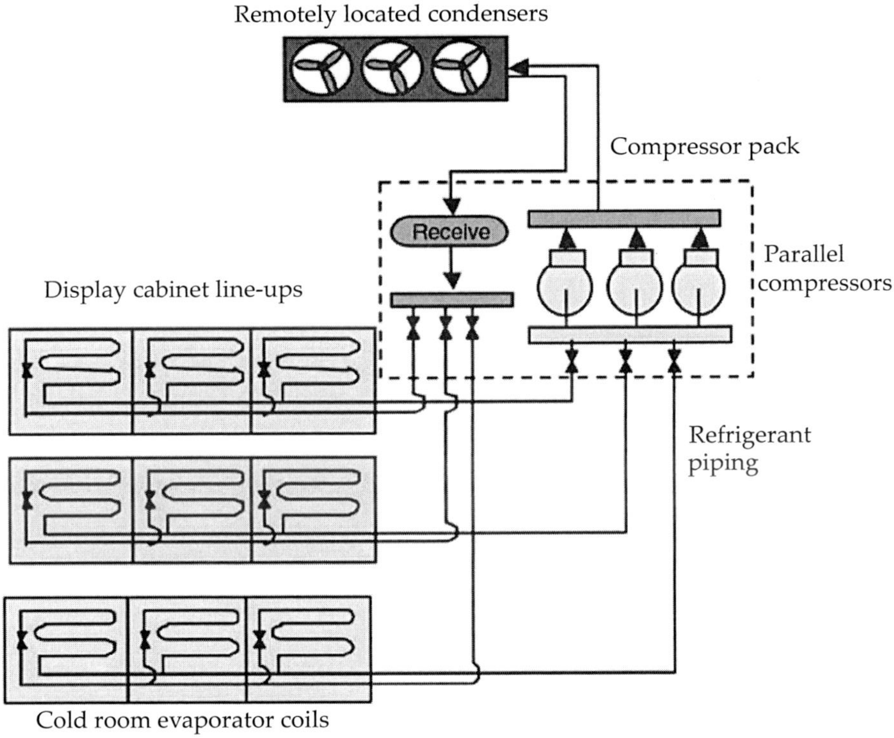

Figure: 3.19 Schematic diagram of a centralized DX system

As mentioned earlier, many food processing industries use a centralized direct expansion (DX) system. These centralized systems offer the flexibility to install compressors and condensers in a central plant area, typically near the store. In the plant room, multiple refrigeration compressors are mounted on compressor packs or racks that also contain the necessary piping, valves, and electrical components needed for operation and control. Evaporators in the cold rooms are supplied with refrigerant from the central plant through distribution pipework installed under the floor or along the ceiling of the sales area. Air-cooled or evaporative-cooling condensers used in conjunction with the multiple compressor systems are generally installed remotely from the compressors, usually on the roof of the plant room.

Separate compressor packs are used for chilling and freezing processes. However, a major disadvantage of a centralized DX system is the large amount of refrigerant required, nearly 4-5 kg/kW of refrigeration capacity. To overcome this issue, a secondary or indirect system is used. In this arrangement, a primary system located in a plant room or on the roof uses natural refrigerants like hydrocarbons or ammonia to cool a secondary fluid, which is circulated to the coils in the display cabinets and cold rooms. Separate refrigeration systems and brine loops are used for the medium- and low-temperature display cabinets and other refrigerated fixtures.

The temperature of individual consumer packs, small individual items, and especially thin-sliced products, which are commonly found in retail display cabinets, responds very quickly to small amounts of added heat. Marketing constraints require that these products have maximum visibility.

4

Operation of Pre-cooling Systems

Pre-cooling is an essential process in the post-harvest stage of perishable produce. It involves swiftly removing heat from freshly harvested fruits and vegetables, reducing quality loss and extending shelf life. The advantages of precooling are evident not only in the enhanced quality of the produce but also in the increased profitability for growers.

In addition to improving quality and shelf life, pre-cooling helps curb the growth of microorganisms. Lower temperatures effectively prevent the growth of both existing and potential microbes, aiding in the preservation of the produce.

Field heat, which refers to the temperature difference between the freshly harvested produce and its optimal storage temperature, highlights the importance of precooling. The method of precooling selected depends on the perishability of the produce and the refrigeration equipment available, emphasizing the need for a customized approach.

Many studies stress the importance of precooling for storing and transporting produce over long distances. In this context, the focus is not on whether to precool, but on how to choose and implement the most effective method. This underlines the critical role of precooling in maintaining the quality, shelf life, and marketability of harvested produce.

Precooling fruits and vegetables is a vital post-harvest procedure aimed at rapidly reducing the produce's temperature to slow down metabolic processes and prolong freshness. The ideal precooling temperature varies based on the type and maturity of the produce but is typically lower than the surrounding temperature. Various methods such as hydro-cooling, vacuum cooling, and forced-air cooling can be used for precooling. Properly understanding and applying these techniques is key to preserving the quality and freshness of fruits and vegetables from the farm to the consumer.

BASICS OF PRE-COOLING

Pre-cooling is a crucial process in the post-harvest handling of fruits and vegetables. It involves subjecting freshly harvested produce to controlled low temperatures before they are transported or stored. This process helps eliminate field heat, slow metabolic activity, and minimize deterioration. By efficiently pre-cooling produce, exporters can extend the shelf life, maintain nutritional quality, and improve marketability.

Moreover, pre-cooling ensures compliance with quality standards, reduces risks associated with temperature fluctuations, and contributes to cost savings and improved customer satisfaction. In essence, understanding the importance of pre-cooling allows exporters to make informed choices, laying the groundwork for long-term success in the global market.

The Importance of Temperature Control

Temperature control is a crucial component of post-harvest handling and storage. Fruits and vegetables are living organisms that continue to respire after harvest. High temperatures can speed up respiration, leading to faster deterioration and a loss of quality. Pre-cooling ensures that the produce is quickly brought to the optimal storage temperature, reducing metabolic activity and helping to extend its shelf life.

Preserving Quality and Freshness

One of the key reasons for pre-cooling fruits and vegetables is to preserve their quality and freshness. Heat stress after harvest can cause various undesirable changes, such as wilting, decay, and loss of texture. Pre-cooling helps maintain the crispness, color, and flavor of the produce, boosting its marketability and consumer appeal. Whether dealing with delicate berries or leafy greens, proper pre-cooling is essential for preserving their visual and sensory qualities.

Reducing the Loss of Moisture

Moisture loss is a common issue during the post-harvest handling of fruits and vegetables. High temperatures can accelerate transpiration, causing dehydration and shriveling of the produce. Pre-cooling helps reduce moisture loss by slowing the rate of transpiration, preserving the firmness and juiciness of the fruits and vegetables. By maintaining appropriate humidity levels, pre-cooling ensures that the produce retains its optimal moisture content during storage and transportation.

Delaying Ripening and Senescence

Many fruits and vegetables continue to ripen after harvest, driven by ethylene production and sensitivity. This is one reason why freeze-dried mango is popular both in India and abroad, as it allows the fruit to be enjoyed year-round. Elevated temperatures can speed up the ripening process, leading to premature softening and decay. Pre-cooling helps delay ripening and senescence by reducing ethylene production and sensitivity. This delay allows for extended storage periods, ensuring that the produce reaches consumers in optimal condition.

Preventing Pathogen Growth

In addition to preserving quality, pre-cooling also plays a vital role in food safety. High temperatures can create ideal conditions for the growth of pathogenic microorganisms, increasing the risk of foodborne illnesses. By quickly lowering the temperature of the produce, pre-cooling slows down the growth of pathogens and reduces the risk of contamination. Proper temperature management is particularly crucial for perishable items such as berries and leafy greens, which are highly vulnerable to microbial spoilage.

Enhancing Nutritional Value

Maintaining the nutritional integrity of fruits and vegetables is essential for ensuring consumer health and wellness. Heat stress after harvest can cause the breakdown of vitamins, antioxidants, and other vital nutrients. Pre-cooling helps preserve the nutritional value of the produce by reducing metabolic activity and minimizing oxidative damage. By keeping the temperatures low, pre-cooling ensures that fruits and vegetables retain their full range of vitamins, minerals, and phytochemicals.

Optimizing Shelf Life

Shelf life is a crucial factor in determining the economic viability of fresh produce. Proper pre-cooling plays a vital role in extending the shelf life of fruits and vegetables, enabling longer storage periods and minimizing losses. By slowing down physiological processes and preventing microbial growth, pre-cooling helps maintain the quality and freshness of the produce throughout the supply chain. As a result, it reduces waste and maximizes profitability for growers, distributors, and retailers.

The Role of Technology in Pre-Cooling

Advancements in refrigeration technology have greatly improved the pre-cooling process, making it more efficient and cost-effective. From forced-air cooling to vacuum cooling, various methods are now available to accommodate different types of produce and operation scales. Automated systems with temperature and humidity control mechanisms allow for precise cooling and monitoring, ensuring optimal conditions for maximum shelf life. Additionally, cold chain logistics solutions enable seamless transportation of pre-cooled produce from farm to consumer, further enhancing food safety and quality assurance.

Pre-cooling is a crucial step in the post-harvest handling of fruits and vegetables. By quickly lowering the temperature of the produce, pre-cooling preserves its quality, freshness, and nutritional value while minimizing moisture loss, delaying ripening, and preventing the growth of pathogens.

With the proper pre-cooling techniques and advanced refrigeration technologies, stakeholders across the supply chain can optimize shelf life, reduce waste, and deliver high-quality produce to consumers globally. Embracing the science of pre-cooling is essential for ensuring the sustainability and profitability of the fresh produce industry in the face of evolving challenges and demands.

Benefits of Pre-Cooling for Perishable Goods Before Export

Pre-cooling perishable goods before export offers several key benefits as follows:

- **Removal of Field Heat:** By eliminating field heat, pre-cooling slows down the natural aging process, thereby extending shelf life.

- **Reduction of Respiration and Ripening:** Pre-cooling slows the rate of respiration and ripening, preserving the freshness of produce during transit.
- **Minimization of Moisture Loss:** It reduces moisture loss, helping to maintain texture and appearance.
- **Prevention of Bruise Damage:** Pre-cooling lowers the risk of bruise damage during transit, ensuring that goods arrive in the best condition.
- **Inhibition of Ethylene Production:** By inhibiting ethylene production, pre-cooling prevents premature ripening.
- **Control of Spoilage Organisms:** It helps inhibit the growth of spoilage organisms, ensuring product safety.
- **Ease on Cooling Systems:** Pre-cooling reduces the burden on refrigeration systems, leading to energy savings and enhanced efficiency.

Pre-cooling is essential for keeping perishable goods fresh, safe, and of high quality during export.

Factors to Consider Before Pre-cooling

Several factors should be considered before pre-cooling perishable goods:

- **Air Temperature During Harvesting**: The ambient air temperature, especially in warmer months, influences the effectiveness of pre-cooling.
- **Time Gap Between Harvest and Pre-cooling:** A longer delay between harvest and pre-cooling accelerates deterioration, making the duration between the two critical.
- **Immediate Pre-cooling for Highly Perishable Crops:** Crops that are highly perishable require immediate pre-cooling to prevent spoilage.
- **Temperature Differential Impact:** The effectiveness of pre-cooling is influenced by the temperature difference between the crop and the cooling medium.
- **Heat Extraction Efficiency:** A crop's ability to release heat efficiently depends on the nature and speed of the cooling medium.
- **Rate of Heat Transfer:** The speed and efficiency of pre-cooling depend on how quickly heat transfers between the crop and the cooling medium.
- **Packaging Considerations:** Using waterproof, ventilated boxes or treating packaging materials with wax can improve air circulation, enhancing the effectiveness of pre-cooling.

These factors play a vital role in ensuring that pre-cooling is done efficiently and optimally.

METHODS FOR PRE-COOLING FRESH PRODUCE

In the post-harvest phase of the perishable produce industry, precooling is an essential step. This procedure quickly removes heat from recently harvested fruits and vegetables, reducing quality loss and increasing product shelf life.

Figure: 4.1 Precooling fresh produce

The main Methods for pre-cooling produce are given below:

- Room Cooling
- Hydro-Cooling
- Vacuum Cooling
- Ice Cooling
- Forced-Air Cooling

Room Cooling

Room Cooling involves placing the product to be cooled in a room with a temperature lower than that of the product for a specified duration until the desired temperature is reached.

There is a trade-off with Room Cooling. While it requires relatively low energy, it is a slow process. This method is best suited for produce varieties that do not decay quickly. Crops such as beets, cabbage, potatoes, pumpkins, apples, pears, and peaches are often room cooled but can benefit from a faster pre-cooling method.

Pre-cooling produce in a cold-storage room or pre-cooling room is a time-tested and widely used method. This method involves placing produce in boxes (wooden, fiberboard, or plastic), bulk containers, or other types of packaging into a cold room where they are exposed to cold air. It is ideal for produce that is sensitive to free moisture or surface moisture. However, room cooling is slow, making it suitable only for small quantities of produce or produce that does not deteriorate quickly.

In this method, cold air is typically discharged into the room near the ceiling and flows past the containers, returning to the heat exchangers. The cold air is usually supplied by forced or induced draft coolers, which use framed, finned evaporator coils with fans to circulate air over the coils. To achieve fast and efficient cooling, it's crucial to use the correct packaging (well-vented) and stacking patterns. The air velocity around the containers should be at least 60 m/min to ensure the necessary turbulence for heat removal.

Since much of the cooling is achieved by conduction, room cooling is slow and results in variable temperature reduction. Therefore, only perishable produce tolerant to slow heat removal should be used. Conventional cold storage is less suited for this method as up to three-quarters of the refrigeration capacity may be required just to remove field heat, with cooling rates often no better than 0.5°C/h. Cold rooms for highly perishable fruits are designed for airflow rates of about 170 to 225 m^3/min for a room with a capacity of 15,000 kg, cooling the produce to 5°C in about 12 hours. Containers should be stacked with proper spacing to allow airflow over the produce and ensure efficient cooling.

To prevent produce from drying out, high relative humidity (90-95%) must be maintained. Containers should be well-ventilated to ensure maximum airflow, and proper spacing between containers, walls, and ceilings is essential. Room cooling is not ideal for bulk bins due to the larger mass of produce. Proper design of the cooling room and refrigeration equipment is vital for efficiency. The refrigeration system should be capable of cooling produce within 24 hours and maintaining the desired storage temperature.

As more commodities are handled in larger quantities and packaged immediately after harvest, room cooling has become more challenging. This slow and variable cooling process extends the cold chain and reduces the product's life during subsequent storage.

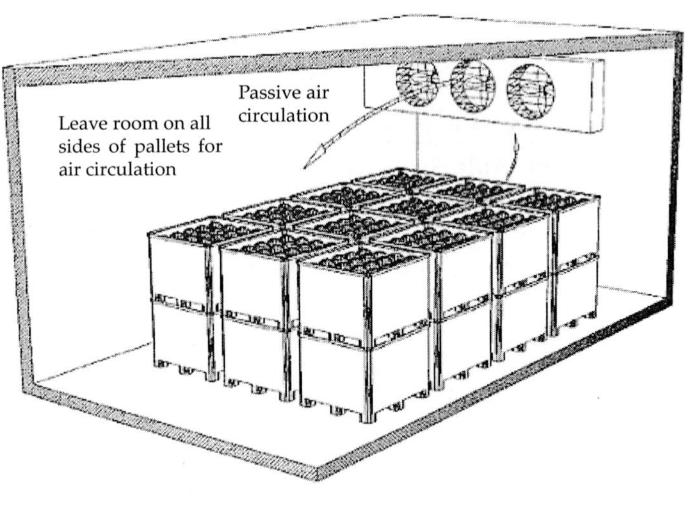

Figure: 4.2 The proper design of the cooling room and refrigeration equipment.

- In room cooling, heat is transferred gradually from the mass of the produce to the cold air circulating around the stacked containers through convection.
- This is the most common and widely used method, where cold air is circulated by fans, cooling the produce through the convection process.
- It is typically used for products with relatively long storage life that are marketed shortly after harvest.
- The advantages of room air cooling include the ability to cool and store produce in the same room, eliminating the need for transfer, thus making it cost-effective.
- In this system, cold air from the evaporator enters the room, moves horizontally, passes through the produce containers, and then returns to the evaporator.

Disadvantage

This system takes longer to cool the products, as the gradual heat removal makes it unsuitable for highly perishable commodities. The products require at least 24 hours to reach the desired storage temperature.

While almost all crops can be cooled using this method, it is most commonly used for citrus fruits, potatoes, onions, garlic, and similar produce.

Hydro-Cooling

Hydro-cooling is a fast precooling method that involves chilling water and either showering or submerging the produce in cold water. While it is an efficient cooling process, it is limited by the temperature that can be achieved in the water. Without additives, water can't be cooled below 32°F (0°C), which makes it ineffective for cooling produce to temperatures below 31°F. Additionally, safety concerns arise, as pathogens can easily spread through the water.

This method is commonly used for cooling asparagus, green beans, broccoli, cabbage, carrots, cucumbers, kohlrabi, parsnips, radishes, rhubarb, rutabagas, and sweet corn.

The principle behind hydro-cooling is that heat transmission from a solid to a liquid is faster than from a solid to a gas, making water a better heat conductor than air.

Cold water is used in this method to quickly cool a wide variety of fruits and vegetables. Hydro-cooling helps avoid moisture loss and may even add water to the produce. The water is usually cooled by mechanical refrigeration, though ice can be used to speed up the process. Additionally, chlorine (150-200 ppm), iodine, nutrients, growth regulators, or fungicides can be added to sanitize, improve nutrient status, and prevent post-harvest diseases.

To achieve quick cooling, cold water must continuously pass over the produce. This can be achieved by submerging the produce in cold water that is continuously circulated through a heat exchanger.

Cooling Time

- 2 minutes for asparagus (long and narrow) and leafy vegetables (which have a larger surface-to-volume ratio).

- 10 minutes for smaller produce like capsicum (large and globular), cherries, and tomatoes.
- Up to 1 hour for larger products such as melons.

Hydro-cooling offers several advantages over other pre-cooling methods, including its ability to clean the produce, provide fast and uniform cooling for commodities, and achieve quicker results than forced air cooling. This method can be performed either by immersing the produce in chilled water or by using a chilled water shower. However, not all crops are suitable for hydro-cooling, as they must be able to tolerate wetting, chlorine exposure, and water infiltration.

Disadvantage

Tank water can become contaminated with microorganisms, leading to higher spoilage rates during subsequent storage or marketing. To prevent this issue, chlorine should be added to the water to ensure its cleanliness and maintain the quality of the produce.

Table: 4.1 The crops which are hydro cooled

Crops Normally Hydro Cooled			
Artichoke	Carrot	Kiwifruit	Potato(early)
Asparagus	Cassava	Kohlrabi	Pomegranate
Beet	Celery	Leek	Radish
Belgian endive	Chinese cabbage	Lima bean	Rhubarb
Broccoli	Cucumber	Orange	Snap beans
Brussels sprouts	Eggplant	Parsley	Spinach
Cantaloupe	Green onions	Parsnip	Summer squash
Cauliflower	J. artichoke	Peas	Sweet corn

Two types of hydro coolers are generally used:

- **Shower/Batch Type:** The water is showered over the commodity, which may be placed in bins, boxes, or moved along a conveyor belt. A typical design involves transporting the crop on a perforated conveyor belt, with the conveyor speed adjustable to match the required cooling time. Cold water is pumped from a tank and sprinkled onto the produce, falling through to the tank below, where it is filtered, recycled, and re-cooled. Efficient cooling relies on a sufficient flow of water over the product surface. For products in bins or boxes, water flow rates of 75-100 liters/min/ft (or 400-600 liters/min/m^2) of surface area are generally used.

- **Immersion type:** This is the simplest type of hydro-cooler, where produce is submerged in cold water. The products, usually in bulk, are in direct contact with the cold water as they move through a long tank. This method works best for products that do not float, as slower cooling would occur if the product simply floated on the water's surface. Immersion hydro-coolers transport the product against the flow of water and often incorporate an agitation system. The water tank should have a depth of at least 30 cm, as water can penetrate inside fruits, especially hollow ones like peppers. The water temperature also plays a role in the level of infiltration. It is recommended that the fruit temperature be at least 5°C lower than the liquid temperature.

This method is suitable for products such as radishes, asparagus, artichokes, green onions, capsicum, and leafy vegetables.

Vacuum Cooling

Vacuum cooling operates by creating a negative pressure using a vacuum pump, which facilitates the evaporation of moisture from the product. This evaporation process cools the product. However, the main drawback of this method is that it leads to some dehydration of the product. Lettuce is one of the most common crops that undergo vacuum cooling. The rapid cooling of horticultural produce is achieved through vacuum cooling by evaporating moisture from the produce.

This evaporation process is enhanced by reducing the pressure to the point where water boils at a low temperature, making the cooling more efficient.

The basic principles of the vacuum cooling process are described as follows:

- At standard atmospheric pressure (1013 mbar), water boils at 100°C. However, this boiling point changes based on saturation pressure. For example, at a pressure of 23.37 mbar, the boiling temperature drops to 20°C, and at 6.09 mbar, it reaches 0°C.
- To transition from a liquid to a vapor state, the latent heat of vaporization must be supplied by the surrounding environment, which helps in lowering the sensible heat of the product.
- The water vapor released from the product during this process must be effectively removed.

Key Components of a 20-Pallet Capacity Vacuum Cooler

In the vacuum cooling process, the pressure inside the vacuum chamber is reduced from atmospheric pressure to about 20 mbar. Initially, during this phase, evaporation occurs slowly, causing minimal cooling, which means the temperature of the produce remains stable until it reaches the saturation pressure at that specific temperature. This point is referred to as the "flash point," where water in the produce begins to vaporize, causing the product to lose moisture and cool down quickly. For example, if the produce has an initial temperature of 20°C, the flash point would occur at 24 mbar. At this point, the wet bulb temperature rises sharply as the air in the tank is evacuated and replaced with evaporated water vapor, which must be quickly removed to maintain an efficient cooling cycle. A condenser within the chamber helps achieve this.

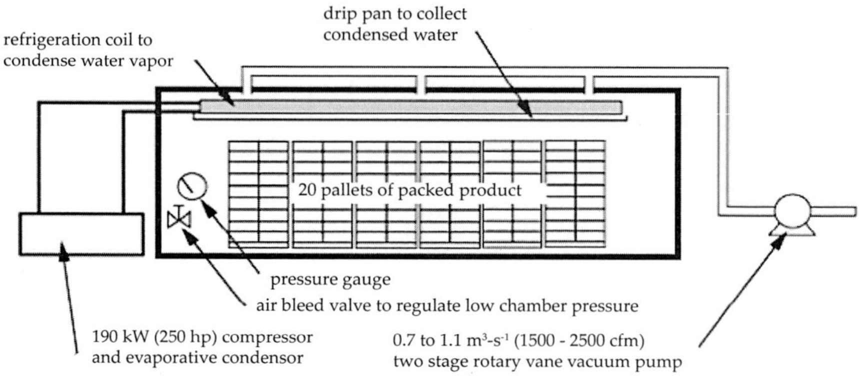

Figure: 4.3 Twenty-pallets of packed product

The pressure is further reduced, and cooling continues until the desired final saturation temperature is reached. In practice, the pressure is generally not reduced below 6.09 mbar (which corresponds to a temperature of 0°C), as further reduction demands extra work and may cause freezing due to the reduced pressure.

The heat required to vaporize the water is taken from the product's surface, which means the cooling rate depends on the heat and mass transfer, particularly the evaporation rate of water from the product's surface and inner tissues. Cooling speed is also influenced by the product's surface-to-volume ratio, the ease with which water is released from the tissues, the vacuum pump's draw rate, and the starting temperature of the produce. Generally, vacuum cooling works best for products with a high surface-to-volume ratio. Since water acts as the primary refrigerant, the quantity of heat removed from the product is directly linked to the amount of water evaporated from its surface.

Vacuum cooling efficiency is greatly influenced by the surface area-to-mass ratio and the ease of water release. Products with low surface-to-mass ratios experience large temperature gradients, which can lead to surface freezing before the product reaches the desired temperature. To prevent this, a "bouncing" procedure is employed, where the vacuum pump is switched on and off to maintain the saturation temperature above freezing. A downside of vacuum cooling is that it causes weight loss in the product due to moisture removal. On average, temperature reductions of 5 to 5.5°C occur for every 1% weight loss, which is generally acceptable for most products. To reduce weight loss, free water may be sprayed on the product before it enters the cooling chamber. Specialized vacuum coolers, known as 'hydrovac' systems, incorporate water sprays near the end of the cooling process to minimize weight loss.

The quick cooling provided by vacuum cooling offers a significant advantage over other cooling methods. As previously mentioned, its ability to rapidly remove heat is one of its key benefits. Additionally, vacuum cooling ensures uniform cooling throughout a package or lot of produce, as long as the package or box is not completely sealed, allowing for free evaporation. This uniformity in cooling helps to maintain the quality of the produce throughout the process.

Vacuum cooling works by removing heat from vegetables through the evaporation of water contained within them. The produce is placed into a sealed container, and air is pumped out, reducing the pressure from the normal air pressure (around 100 KPa) to a near-vacuum level (<1 KPa). In these conditions, water boils at temperatures below 7°C. As the water in the vegetables turns from liquid to gas, it absorbs heat, which cools the product. This vapor is then drawn past refrigeration coils, where it condenses back into liquid water.

For vacuum cooling to be effective, the vegetables need to lose moisture easily, which is why this method is ideal for leafy produce like lettuces, Asian greens, and silverbeet. It is also effective for products like broccoli, celery, and sweet corn. However, vacuum cooling is not suitable for items with waxy skins or low surface area to volume ratios, such as carrots, potatoes, or zucchini.

To achieve a 5°C reduction in temperature, approximately 1% of the produce's weight must be converted into water vapor. Modern hydro-vacuum coolers help minimize moisture loss by spraying water over the produce during the cooling process, significantly reducing the loss to almost negligible levels.

Figure: 4.4 Hydro-vacuum Cooler Loaded with bins of broccoli

Vacuum cooling is the quickest cooling method for suitable products, typically requiring only from 20 to 40 minutes to reduce the temperature of leafy vegetables from 30°C to 4°C. For example, vacuum cooling reduced the temperature of harvested broccoli by 11°C in just 15 minutes. Large vacuum coolers can cool multiple pallets or bins simultaneously, easing the load on cool room systems. This process can even be applied to packed cartons, provided there is adequate venting to allow air and water vapor to escape efficiently.

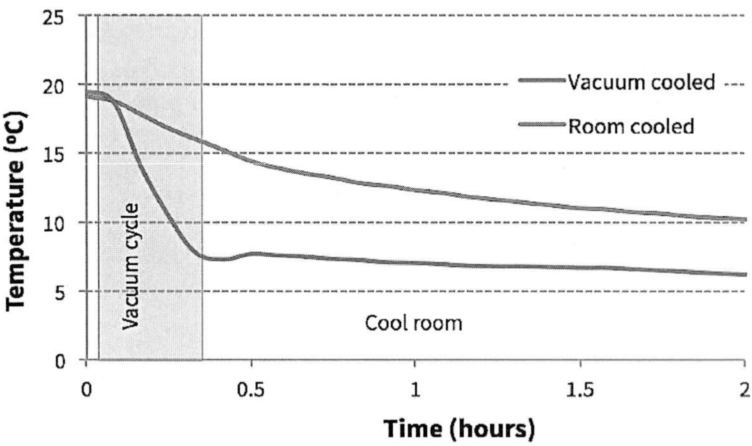

Figure: 4.5 Temperatures of broccoli vacuum cooled then transferred to the cool room or simply placed in a plastic crate in the cool room. The vacuum cooler reduced the temperature of broccoli from 18.8°C to 7.8°C in 15 minutes.

Vacuum cooling stands out as the most energy-efficient cooling method, with nearly all the electricity used contributing directly to lowering the temperature of the product. Since the unit is sealed during operation, there are no factors like lights, forklifts, or workers that could inadvertently increase the temperature. Additionally, vacuum coolers prevent any infiltration during the cooling process.

Hydro-vacuum coolers, though slightly less efficient, also cool the water in the system. For every 400-carton load, approximately 570L of cold water are lost, requiring replacement and re-cooling. Additionally, cold water on the sides of the unit can cool the walls, leading to some heat transfer into the unit. However, these losses are relatively minor compared to those in most cool rooms.

Cold Wall

A permanent false wall or air plenum contains an exhaust fan that draws air from the room and directs it to the cooling surface. The wall is at the same end of the cold room as the cooling surface. The wall is constructed with a damper system that only opens when containers with openings are placed in front of it. The fan draws air from the cold room through the container and its contents, cooling the products.

Ice Cooling

Ice cooling, or ice injection cooling, involves mixing ice and water into a slurry, which is then "injected" into the product packaging to cool it by direct contact. Historically, crushed ice was blown onto the tops of railcar loads of lettuce, and this method is still used for some produce varieties such as asparagus, broccoli, kohlrabi, leeks, onions, parsnips, radishes, rutabagas, and sweet corn.

Before the advent of refrigerated trucks and cool rooms, ice was commonly used to cool produce during transport. Although, it is less common now, ice is still used for some products, like broccoli and Brussels sprouts, for several reasons: it keeps the produce cold and hydrated, appears visually appealing when the carton is opened, and meets customer expectations.

Ice can often be colder than 0°C, as freezers that make ice can operate well below freezing, sometimes near -20°C. Even robust products like broccoli and Brussels sprouts can suffer freezing damage when exposed to ice this cold. As ice melts, it absorbs energy from the surroundings, cooling the produce. However, if the ice is still solid when the carton is opened, it has not been used effectively.

Once the ice has melted, the produce will be wet, which can lead to issues like splitting, discoloration, and increased rotting or disease. To prevent contamination, the water used for icing must meet high microbial standards, free from pathogens, fungi, and bacteria that can cause plant diseases.

Although, ice cooling is effective, it also has drawbacks: it requires significant energy to make the ice, adds an extra packing step, and increases transport costs due to the added weight and volume. Additionally, the water that melts from the ice can lead to undesirable conditions for the produce.

Figure: 4.6 Broccoli is often packed in ice.

Figure: 4.7 Ice can cause freezing damage to broccoli florets (left). If the ice melts, broccoli left sitting in liquid water is likely to split and develop rots (right).

Air Cooling by Force

Forced-air cooling is a process where heat is removed by generating a pressure differential across the product, which causes cold air to be drawn through the packaging of the product. This technique is also referred to as blast cooling. Various types of produce, such as apples, can be effectively cooled using this method:

- Apples
- Apricots
- Avocados
- Beans (green, wax)
- Bell peppers
- Blueberries
- Boysenberries
- Broccoli
- Brussel sprouts
- Cabbage
- Carrots
- Cantaloupe
- Cauliflower
- Celery
- Chili peppers
- Cucumbers
- Figs
- Grapes (table)
- Honeydew
- Kiwifruit
- Kohlrabi
- Leaf lettuce
- Mushrooms
- Nectarines
- Olives
- Peaches
- Peas
- Pears
- Peppers
- Plums
- Potatoes
- Raspberries
- Rhubarb
- Romaine lettuce
- Saskatoons
- Spinach
- Squash
- Strawberries
- Swiss chard
- Tomatoes

Figure: 4.8 Forced-air cooling

- In this system, cold air is forced through the product from one side to the other using a large fan.
- The cold air moves through the containers instead of around them.
- The cooling process is 4 to 10 times faster than room cooling, and the rate depends on the airflow and the volume of the produce.
- The air is blown at a high velocity, which can lead to desiccation of the produce. To mitigate this effect, cold water sprays are used to blow air through the product.
- Although slower than hydro-cooling, this method is an excellent alternative for crops that require rapid heat removal but cannot tolerate wetting or chlorine in the cooling water.
- Sufficient airflow is essential since fruits in the center of packages tend to lose heat more slowly than those on the exterior.
- This system is also known as a high humidifier. A relative humidity (RH) of 90-95% must be maintained in the pre-cooler to prevent dehydration during the cooling process.
- This system is applicable to a wide range of crops, including berries, ripe tomatoes, bell peppers, various fruits, cabbage, green peas, cucumbers, brinjal, muskmelon, watermelon, and mushrooms.
    - **For baby corn:** 5-6 hours of cooling at 2-4°C.
    - **For leafy vegetables:** 1-2 hours of cooling at 6-8°C (note that too low temperatures can cause chilling injury).

Forced air, also known as "pressure cooling," enhances the cooling process by increasing the surface area that is being cooled, extending from just the package to the produce inside. This method can reduce cooling times by as much as 10 times compared to room cooling.

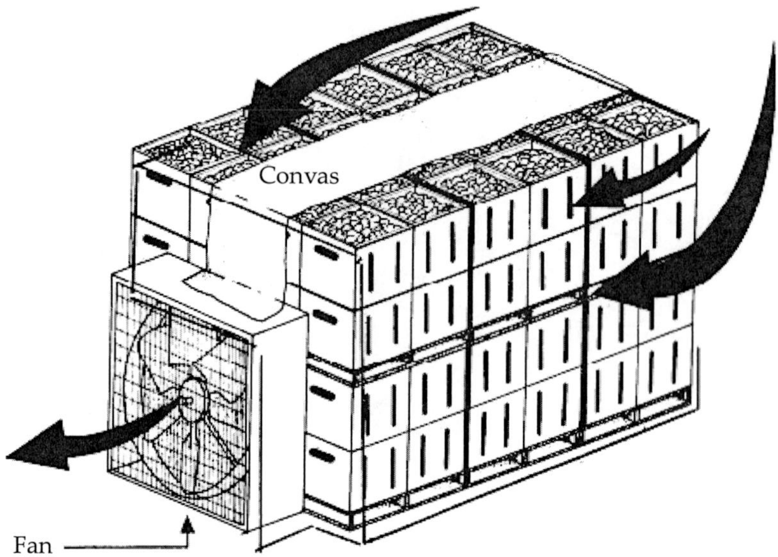

Figure: 4.9 Tunnel-type forced air cooling using the cold wall system

In forced air systems, cold air is pulled through vented packages at rates ranging from 0.1 to 2.0 L/second/kg. A general guideline for fan strength is that it should be powerful enough to hold a piece of A4 paper against one of the carton vents.

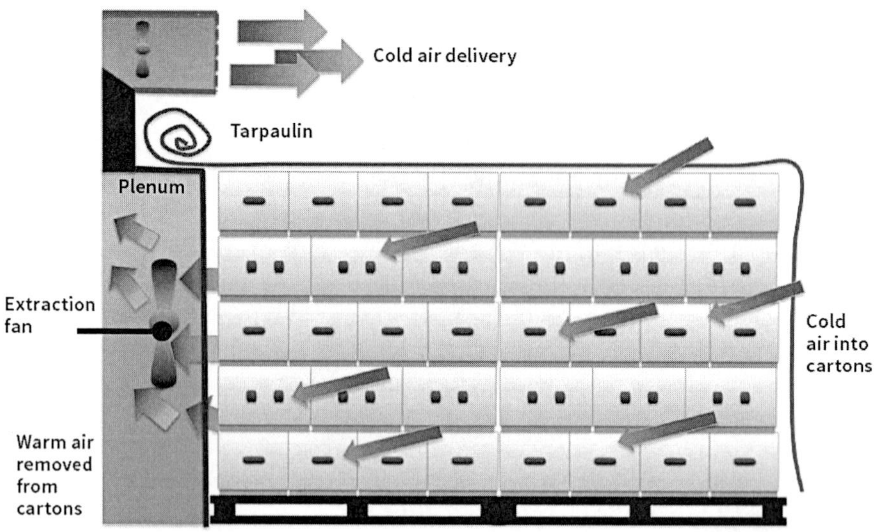

Figure: 4.10 Forced air-cooling system for large volumes of packed cartons or bins.

These systems are typically designed for two rows of stacked pallets (or bins) placed against a central plenum. A tarpaulin is draped over the top to seal the gaps between the pallets, directing air through the carton side vents.

The fan inside the plenum draws air through the cartons, removing heat from the packed produce. This air may either be exhausted directly back into the room or pass through a cooling system before being released.

For forced air cooling to function effectively, cartons should have vents covering at least 5% of their surface area at both the entry and exit points. The vents must be aligned between cartons, even when pallets are cross-stacked. Additionally, one or two large holes allow for better air movement than multiple smaller holes, even if their total area is the same, due to the edge effect around the vents.

For smaller product quantities, a basic bin cooling system can be utilized. This involves placing a fan, enclosed in housing, on top of one or more vented half-tonne bins. The vents on the side of the bins are blocked using plastic wrap or tarpaulin, directing air to move exclusively through the base of the bins. The air is then drawn through the bins and exhausted directly into the room. This system is effective for cooling smaller volumes of produce, particularly when cool room space is limited.

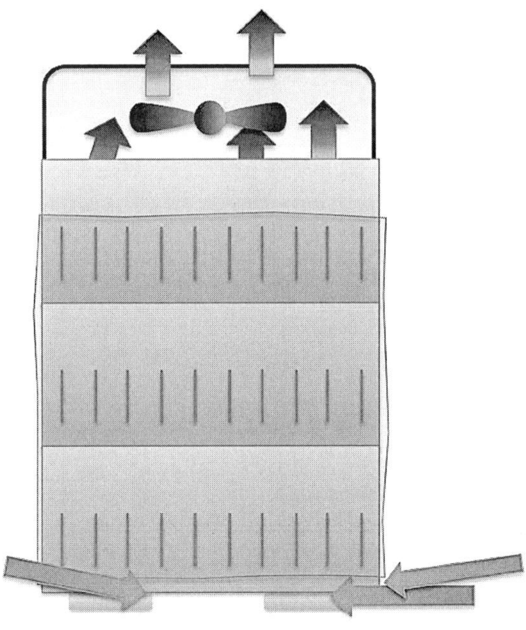

Figure: 4.11 Forced air-cooling system for small volumes of bins.

Moisture loss typically isn't a concern during forced air-cooling due to the fast nature of the process. However, high-humidity systems can be utilized if moisture loss becomes an issue. Unlike room cooling, condensation doesn't occur with forced air systems. Condensation can promote disease and weaken cardboard packaging. In the absence of positive air movement, water vapor released by warm produce may condense on the cold product or packaging near the air delivery system. With forced air systems, as the air passes through the produce, it warms up, enhancing its ability to hold and carry away water vapor, thus preventing condensation. The energy efficiency of forced air systems can vary significantly.

In some cases, forced air rooms are also used for storage, which can reduce efficiency, particularly if fans are left on between cooling cycles to maintain the room's temperature. Fan efficiency can also contribute to variability in performance.

The Distinction Between Forced Air Cooling and Vacuum Cooling

Forced Air Cooling

The air passes over the surface of the crop, cooling the outside while the inside is cooled by heat transfer from inside to the outside for the crop.

Vacuum Cooling

In cooling chamber, pressure (reduced) is exactly the same around the produce and in the centre of the produce. This means the cooling is very even and quick throughout the crop.

Table: 4.2 The main crops which are pre-cooled by forced air.

Crops Usually Pre-cooled by Forced Air			
Anona	Citrus	Litchi	Plantain
Avocado	Coconut	Mango	Pomegranate
Banana	Cucumber	Mangosteen	Prickly pear
Barbados cherry	Eggplant	Melons	Pumpkin
Berries	Fig	Mushrooms	Rhubarb
Breadfruit	Ginger	Okra	Sapota
Brussels sprouts	Grape	Orange	Snap beans
Cactus leaves	Grapefruit	Papaya	Strawberry
Capsicum	Guava	Passion fruit	Summer squash
Carambola	Kiwifruit	Persimmon	Tomato
Cassava	Kumquat	Peas	Tree tomato
Cherimoya	Lima bean	Pineapple	Yam

Serpentine Cooling

A serpentine system is designed for cooling bulk bins and is a variation of the cold-wall method. The bulk bins, which feature vented bottoms (with or without side ventilation), are stacked several units high and deep, with the forklift openings positioned against the cold wall. Every other forklift opening in the stack is sealed with canvas, while the alternate unsealed openings allow cold air to flow through the produce. Cold room air is drawn through the produce via the unsealed openings in the stack and the top of the bin, facilitating efficient air circulation for cooling.

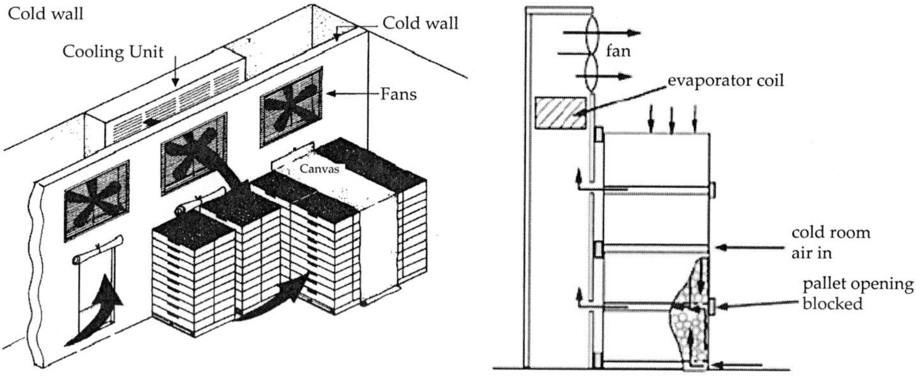

Figure: 4.12 Serpentine cooling

Forced air cooling is much faster than conventional room cooling because the cooling air comes into direct contact with the product being cooled. Typically, forced air cooling is 4 to 10 times faster than room cooling, but methods like hydro-cooling and vacuum cooling can be 2 to 23 times faster than forced air cooling. One advantage of forced air cooling is that it can often be easily and inexpensively implemented by converting existing facilities, as long as there is adequate refrigeration capacity and cooling surfaces.

However, when extremely rapid cooling is necessary, forced air cooling may be more costly compared to other methods, which can limit its use for products requiring quick cooling. Additionally, forced air cooling relies on a specific stacking pattern, meaning skilled operators are needed to ensure the correct loading arrangement for optimal cooling rates.

Hydrocooling

Hydrocooling is a process that uses chilled or cold water to lower the temperature of produce, whether in bulk or smaller containers, before packing it for further transport. This method involves either flooding, spraying, or immersing the product in cold water. There are various types of hydrocoolers in commercial use, each differing in cooling efficiency and rates.

The distinctions between the methods lie in how the cooling is achieved and how the produce is moved or positioned within the cooler. Some common types of hydrocoolers include the conventional (flood) type, immersion type, and batch type. In a flood-type hydrocooler, the product is cooled by flooding as it moves through a cooling tunnel. In the batch system, chilled water is sprayed over the product for a set period based on the season and initial product temperature.

These batch systems are less expensive because of their smaller capacity, though they are prone to uneven cooling, leaving "hot spots" throughout the product load. For hydrocooling to be effective, it is essential that the water evenly contacts the product's surface, but this is not always achieved with conventional or batch systems, leading to undercooling in some areas.

The bulk or immersion-type cooler combines both immersion and flood cooling. The produce is immersed in cold water until it is gradually lifted by an inclined conveyor, passing through an overhead shower. This method delivers the fastest hydrocooling, nearly doubling the speed of conventional systems, as the moving chilled water surrounds the product, speeding up temperature reduction. Additionally, the bulk cooler allows for greater flexibility, as it supports packaging after cooling, making it more suitable for a broader range of growers.

Hydrocooling is most effective for produce that can withstand wetting and not be damaged by water or disinfectants, making it ideal for crops requiring washing before market preparation. A challenge with many hydrocoolers is the risk of contamination from recirculated water, which can lead to the accumulation of decay-causing organisms.

To mitigate this, mild disinfectants like chlorine (at 100 ppm) or approved phenol compounds are added to the water. A key benefit of hydrocooling is its ability to prevent moisture loss during cooling, along with its rapid speed compared to other methods. Field heat can be removed in just 20-30 minutes with hydrocooling, whereas forced air-cooling may take several hours. Hydro-air cooling, a specialized form of hydrocooling, uses a mixture of refrigerated air and water mist that is circulated around the produce. This technique reduces water usage and improves sanitation. The air-to-water ratio in hydro-air cooling affects its heat transfer capabilities, determining which products are best suited for this method.

Shower coolers operate by distributing cold water from a refrigeration evaporator through a perforated metal pan. These coolers can either function with a continuous flow using a moving conveyor or operate in a batch mode. Immersion coolers, on the other hand, are designed for products that sink in water, but they generally cool slower than shower coolers due to the lower water flow rate over the produce.

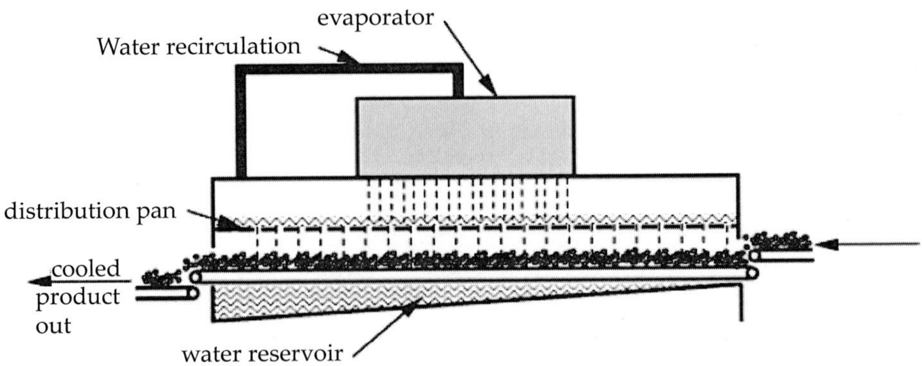

Figure: 4.13 Cut-away side view of a continuous-flow shower-type hydrocooler

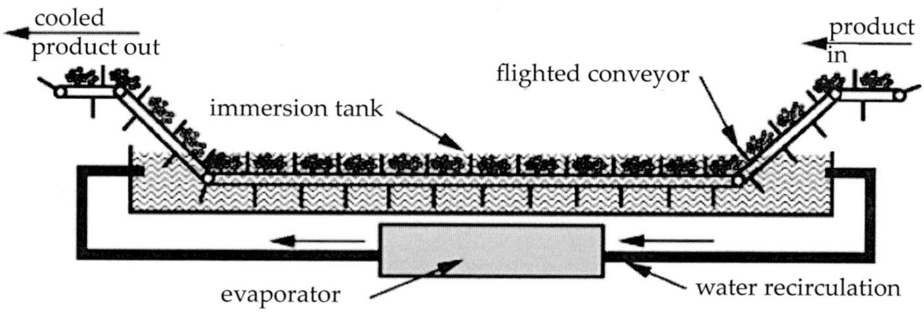

Figure: 4.14 Cut-away side view of a continuous-flow immersion hydrocooler

Water, being a better heat transfer medium than air, makes hydrocooling much faster than forced-air cooling. In properly designed shower coolers, small products such as cherries can be cooled in under 10 minutes, while larger items like melons take approximately 45 to 60 minutes to cool. Immersion coolers typically take longer to cool than shower coolers due to slower water movement past the produce.

Packaging used in hydrocooling must allow water to flow vertically through it and be durable enough to withstand water contact. Materials such as plastic or wood containers are ideal, while corrugated fiberboard requires wax dipping to resist water damage. One advantage of hydro-cooling is that it prevents moisture loss during the cooling process and can even help rehydrate slightly wilted produce.

Cryogenic Cooling

Cryogenic precooling utilizes the latent heat of evaporation from liquid nitrogen or solid CO_2 (dry ice), which generates extremely low boiling points of -196°C and -78°C, respectively. In this process, produce is cooled as it passes through a tunnel where liquid nitrogen or dry ice evaporates. However, these low temperatures can freeze the produce, ruining it for fresh market purposes. To prevent this, the evaporation rate and conveyor speed are carefully controlled.

While cryogenic cooling is inexpensive to install, it is costly to operate. This method is primarily used to cool crops like soft fruits, which have a short seasonal production period. By using cryogenic cooling, growers can avoid the high capital costs of other cooling techniques during the season of use. However, the high costs of liquid nitrogen, dry ice, and other suitable non-toxic refrigerants make this technique most appropriate for cooling higher-value products.

Evaporative Cooling

Evaporative cooling is a cost-effective method for reducing the temperature of produce, particularly effective in areas with low humidity. In this method, dry air is passed through moist padding or a fine mist of water, then through containers of produce that are vented. As the water evaporates, it absorbs heat from the surrounding air, thus lowering the temperature of the produce. For optimal results, the incoming air should have a relative humidity of less than 65 percent.

Typically, this cooling technique can lower the temperature by 10-15°F. It is particularly suitable for warm-season crops that require warmer storage temperatures, such as tomatoes, peppers, cucumbers, and eggplants, which typically need storage temperatures of 45-55°F.

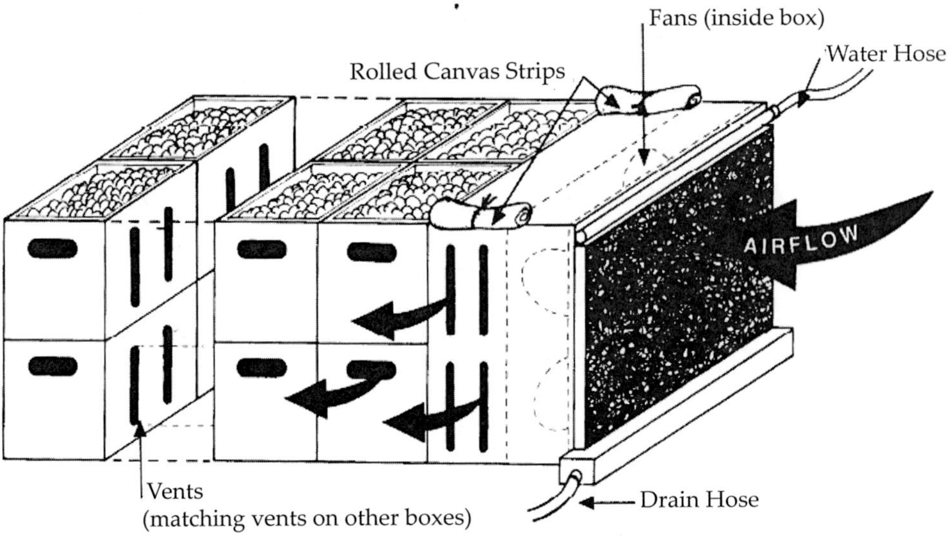

Figure: 4.15 Evaporative cooling

CONSEQUENCES OF SKIPPING PRE-COOLING FOR PERISHABLE SHIPMENTS

Neglecting to pre-cool fruits and vegetables before export can lead to severe repercussions, including:

- **Increased Food Waste:** Without proper pre-cooling, produce becomes more vulnerable to damage during transport, leading to greater food waste and the loss of valuable resources.
- **Rejection by Buyers:** Buyers may refuse the shipment due to quality problems like wilting, discoloration, or spoilage, which can result from inadequate cooling. This can harm the exporter's reputation and jeopardize future business.
- **Claim Rejection by Shipping Lines:** Shipping lines may refuse to honor claims for damaged goods if there is no proof of pre-cooling, which can lead to significant financial losses for the exporter.
- **Profit Loss:** Besides the cost of wasted produce and rejected shipments, exporters may experience a reduction in profits and a diminished return on investment due to disappointed customers and lost sales opportunities.
- **Loss of Trust:** Failing to deliver high-quality produce due to improper pre-cooling can damage buyers' trust in the exporter's reliability and professionalism, leading to the end of long-term business relationships.

Mechanism of Pre-Cooling

Conduction and convection are the primary methods for transferring heat during the cooling process of produce. Conduction involves heat moving directly within the produce to its coolest surface, effectively transferring heat from one object to another (e.g., from fresh produce to water or from a warmer to a cooler surface). In contrast, convection transfers heat away from the surface of the produce through a cooling medium such as moving air or water.

Produce such as potatoes, apples, cauliflower, oranges, and other fruits with larger mass and smaller surface area require more time to pre-cool compared to items with smaller mass and larger surface area, like lettuce, green onions, carrot tops, peas, corn, and brussel sprouts. This is because in larger produce, the heat from the interior must first move to the surface before it can be transferred out.

The cooling rate is influenced by both the volume and the exposed surface area of the product. Additionally, the temperature difference between the product and the cooling medium must be considered. For instance, leafy vegetables with large exposed surfaces cool approximately five times faster than large fruits such as melons, which have more volume but less surface area.

5

Operation & Maintenance of Cold Storage and Solar Cold Room

Cold storage is a widely used method for handling perishable goods in bulk between production and marketing stages. It is essential for preserving the freshness and quality of perishable commodities by controlling temperature and humidity within the storage environment. Ensuring the temperature remains sufficiently low is crucial, as failing to do so can lead to chilling damage to the produce. Additionally, maintaining a relative humidity of 85-90 percent in the storage area is vital for most perishables, as lower humidity can negatively affect their quality.

Fruits and vegetables, if stored at ambient harvest temperatures, have a very short shelf life. Post-harvest cooling helps to quickly remove field heat, thus extending the storage duration. Proper cooling after harvest can:

- Reduce respiration and enzymatic degradation;
- Minimize water loss and prevent wilting;
- Slow or stop the growth of microorganisms that cause decay;
- Lower the production of ethylene, the natural ripening agent.

Besides preserving quality, post-harvest cooling offers greater marketing flexibility by allowing growers to sell produce at the ideal time. With cooling and storage facilities, growers can delay marketing and avoid the need for immediate sale after harvest. This flexibility is beneficial for suppliers to restaurants, grocery stores, or small-scale growers seeking to consolidate shipments. Post-harvest cooling is essential for delivering the highest quality produce to consumers.

Cold storage can be enhanced with additional elements like carbon dioxide, sulfur dioxide (for grapes), or nitrogen, depending on the product's nature. The storage of dried or dehydrated vegetables can also be efficiently managed for over a year at temperatures between 0°C and 10°C, with relative humidity levels of 80-95 percent.

In recent years, cold storage technology has made significant progress, leading to improved preservation of organoleptic qualities, reduced spoilage, and longer shelf life. These advancements are the result of collaborative efforts between physiologists, who identify the needs of various fruits and vegetables, and refrigeration experts who design and optimize cooling systems accordingly.

ROOMS FOR COLD STORAGE

Cold storage refers to a specialized facility where the temperature is maintained at very low levels using advanced machinery and precise instruments. These storage units are considered a vital part of the infrastructure for perishable goods, including fruits, vegetables, seafood, meat, dairy products, and more. Cold storage is crucial for prolonging the shelf life of these products, preserving their quality, and preventing or delaying spoilage. From the viewpoint of farmers, traders, and consumers, cold storage plays an important role in preventing surplus stock, stabilizing market prices, and ensuring consistent pricing across both demand fluctuations and time periods. In this way, cold storage rooms are instrumental in minimizing the loss of perishable goods.

However, these facilities must also address operational challenges, such as maintaining the integrity of the cold chain throughout the storage process, minimizing energy consumption, and ensuring traceability of the products.

Frozen Cold Storage

Frozen cold storage, often referred to as a frozen warehouse, consists of specialized facilities built to maintain consistently low temperatures, typically below freezing, for storing frozen food products. These storage units are essential for preserving the safety and quality of frozen foods throughout the entire supply chain.

Frozen cold rooms are engineered to maintain temperatures generally ranging from -18°C (0°F) to as low as -30°C (-22°F). Additionally, the relative humidity in these facilities is kept as high as possible within the limits of the available equipment and effective operational practices. This high humidity level helps to prevent issues such as "freezer burn" and moisture loss, especially when the food packaging materials may not provide adequate protection.

Comparison Between Cold Storage and Frozen Cold Storage

Below are the prime differences between cold storage and frozen cold storage:

Temperature Range

Cold storage rooms are designed to maintain temperatures ranging from approximately 0°C (32°F) to 15°C. This temperature range is ideal for storing perishable goods and produce that require cooling but do not need freezing conditions.

In contrast, frozen cold storage facilities operate at much lower temperatures, typically ranging from -18°C (0°F) to as low as -30°C (-22°F). These extremely low temperatures are essential for the long-term storage of frozen food products and pharmaceutical items, ensuring their preservation over extended periods.

Commodity Type

Cold storage facilities are specifically designed to prolong the shelf life of perishable products that do not require freezing. These include dairy products, bakery goods, fresh produce (fruits and vegetables), processed foods, seafood, and meats.

On the other hand, frozen cold storage is used to store and preserve frozen goods, such as frozen vegetables and fruits, frozen meats, ice cream, pre-cooked meals, frozen desserts, and similar items.

THE VITAL ROLE OF COLD STORAGE IN FOOD SAFETY

Food safety is of utmost importance in safeguarding the health and well-being of individuals and communities across the globe. Perishable products such as fruits, vegetables, dairy items, and meats are especially vulnerable to spoilage and contamination by harmful bacteria. Cold storage becomes a vital protective measure, serving a crucial role in ensuring food safety across the entire food supply chain, from farm to fork.

Understanding the Science Behind Food Spoilage

Food spoilage is mainly caused by the growth of microorganisms, such as bacteria, molds, and yeasts, which thrive in warm conditions, multiply quickly, and consume the food's nutrients for their own growth. This not only diminishes the food's nutritional value and quality but also leads to the production of toxins that can cause foodborne illnesses.

Cold storage helps by slowing the metabolic activity of these microorganisms. By maintaining a low temperature, it significantly inhibits the enzymatic processes responsible for their growth and reproduction, thereby extending the shelf life of perishable foods and ensuring they remain safe and edible for a longer duration.

The Advantages of Cold Storage in Food Safety

The implementation of cold storage practices offers a multitude of advantages for food safety, which are as follows:

- **Reduced Bacterial Growth:** Cold temperatures play a crucial role in slowing down the growth of harmful bacteria that can lead to foodborne illnesses. This helps minimize the risk of contamination, ensuring consumer safety.
- **Preserved Nutritional Value:** By maintaining low temperatures, cold storage preserves the essential vitamins, minerals, and other nutrients in food. This ensures that consumers benefit from the full nutritional value of the products they consume.
- **Extended Shelf Life:** Cold storage helps delay spoilage, keeping food fresh and edible for a longer time. This not only reduces food waste but also ensures that resources are used optimally.
- **Improved Food Quality:** Cold storage helps retain the texture, flavor, and overall quality of food. This enhances the consumer experience and reduces potential economic losses from spoilage-related issues.

ESSENTIALS OF DESIGNING A COLD STORAGE PROJECT

The design of cold storage facilities is typically aimed at ensuring the proper storage of perishable goods at designated temperatures, while also considering a balance between initial investment, operating expenses, maintenance costs, and depreciation. According to the guidelines provided by the MIDH, the projects are recommended based on the cost components for each phase.

Table: 5.1 The cost components for the phases of cold storage projects

Item	% of the project cost (range)
Civl construction	50-55
Thermal insulation	10-15
Refrigeration system	20-25
Electrical system	10-15

The fundamental procedures for constructing cold storage units should include the following essential requirements.

Process Layout

The primary consideration for any food project involving insulated enclosures is to first determine the layout of the operation that will be contained within the structure. For example, in a meat plant, this could involve a carcass dressing line or a boning room. Similarly, for a cold storage facility, the pallet layout and the mode of operation must be defined upfront. It is ineffective to construct an envelope and then try to fit the processing machinery into it afterward.

Planning Drawings and Application

Only after finalizing the process layout can a planning application be submitted, as this will allow the dimensions of the envelope and supporting buildings to be determined and finalized.

Design Drawings and Specifications

After obtaining planning approval, the preparation of design drawings and specifications can begin. For a competitive design-and-construct tender, it is essential to prepare approximately 15–20 detailed drawings, which should include, at a minimum, the process layout, elevations and sections, refrigeration system layout, mechanical and electrical systems distribution, and lighting layout.

Additionally, to complete the package, at least six separate detailed specifications are necessary, outlining the project's requirements on:

- Contractual requirements
- Building specification
- Refrigeration specification
- Insulation panel supply and erection
- Electrical requirements
- Mechanical services.

The chosen location for the cooling facility should align with its main purpose. If the facility is intended for retail sales of fresh produce, it should be conveniently located with easy access to public roads. However, if the primary role of the cooling facility is to cool and assemble wholesale lots, public accessibility becomes less important.

In such cases, the best location might be next to the packing or grading room. This space could house grading and packing equipment, as well as store empty containers and other supplies when not in use for cooling. Regardless of its specific use, the facility will need reliable access to electrical power and water. For larger cooling rooms requiring more than 10 tons of refrigeration in a single unit, three-phase power will be necessary. It's crucial to consider the location of existing utility lines, as connection costs can be significant, especially in rural areas. Consulting the local power company for details is advisable.

Additionally, when planning the location and design of the facility, it's wise to consider potential future growth. A refrigerated store, equipped with thermally insulated spaces and refrigeration machinery, should be designed to ensure the following services:

- The products nature.
- Frequency of loading and unloading.
- Harvest and dispatch schedule.
- Field heat of the produce.
- Daily volume of produce to be handled.
- Daily volume of ice to be produced.
- Package dimensions and nature.

The availability of both skilled and unskilled labor from the local area is a crucial factor to consider for the successful operation of the facility.

Figure: 5.1 Cool recovery system during air exchange and proper handling at cold store.

Equipment Selection

Once the hourly BTU load has been calculated, equipment selection can proceed. Some of the key factors influencing the choice of equipment include:

- Equipment Balance
- Temperature Difference (TD)
- Product Safety or Capacity Control
- Operation Type or Air Flow

Figure: 5.2 Few additions like air curtain, strong pallet &pre-cooling facility that improve the quality of stored products.

Equipment Balance

The condensing unit is typically chosen first, ensuring it has a capacity that exceeds the calculated cooling or freezing load. The unit cooler(s) should then be selected to match the capacity of the condensing unit. The condensing unit's capacity must be chosen based on a suction temperature (after adjusting for suction line pressure drop) that aligns with the unit cooler(s) and maintains an optimal temperature difference (TD) between the refrigerant in the unit cooler and the air in the refrigerated storage space. Additionally, the condensing unit's capacity should be selected to accommodate the condensing temperature related to the available condensing medium (ambient air or water) at the installation site.

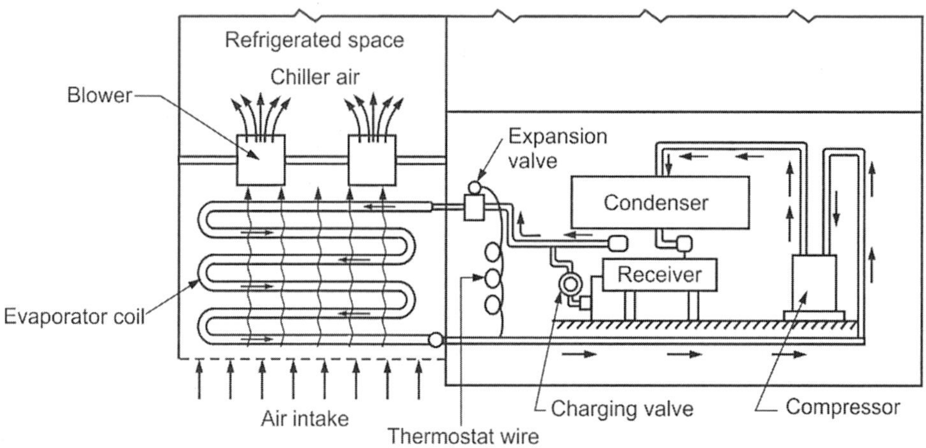

Figure: 5.3 Schematic description of refrigeration system in a typical cold store.

TECHNOLOGICAL DEVELOPMENTS AND INNOVATIONS IN COLD STORAGE

Recent innovations in cold storage technology have transformed the way fruits and vegetables are preserved. Cryogenic freezing, which uses liquid nitrogen for rapid cooling, has become a game-changer by minimizing cellular damage. The latest advancements in compressor technology have made systems more energy-efficient, offering improved temperature control and lowering operational costs.

Real-time monitoring systems are now in place to maintain stable environmental conditions, with IoT devices providing remote management and data analysis. The integration of robotics and automation in storage management has improved precision, reduced labor costs, and streamlined operations.

Moreover, the increasing use of eco-friendly refrigerants is helping lower the carbon footprint of cold storage facilities. These technological breakthroughs result in longer shelf life and superior quality of stored produce.

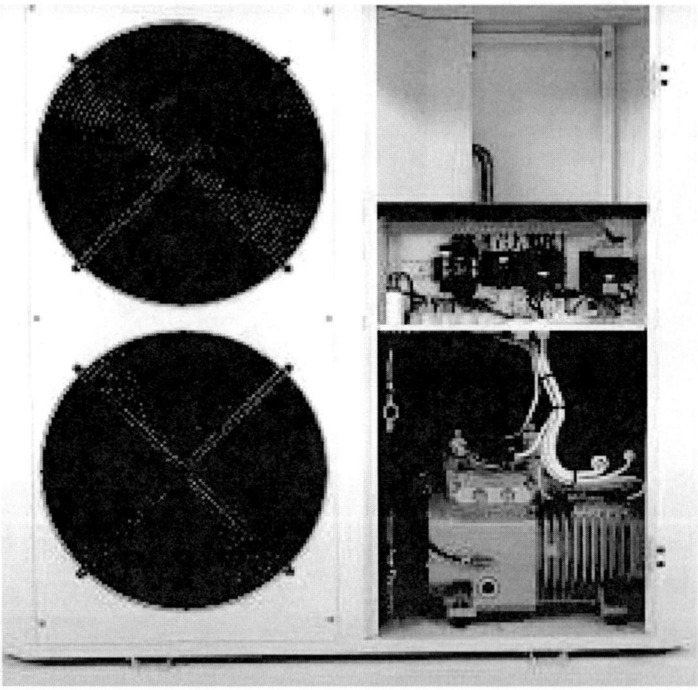

Figure: 5.4 Cold storage cooling unit

Hygiene and Safety Procedures in Cold Storage Environments

Ensuring hygiene and safety in cold storage units is essential for maintaining the quality of fruits and vegetables. Workers must follow strict guidelines to ensure safe operations:

- Adhering to personal hygiene practices, such as frequent hand washing and wearing clean uniforms, helps prevent contamination.
- Cold storage areas should be regularly cleaned and sanitized to prevent bacterial growth and cross-contamination.
- Workers must wear safety gear like insulated gloves, non-slip shoes, and proper cold-weather clothing to protect themselves from cold-related injuries.

- Proper training in handling equipment minimizes accident risks and promotes efficient operations.
- Regular monitoring for gas leaks, particularly ammonia from refrigeration systems, ensures the well-being of employees.
- Implementing a HACCP (Hazard Analysis and Critical Control Points) plan helps manage potential biological, chemical, or physical hazards.

These practices are essential to preserve the quality of produce and maintain a safe and healthy work environment.

Tips for Consumers on Cold Storage: Increasing the Shelf Life of Produce at Home

Here are some tips for maintaining the freshness of fruits and vegetables in your refrigerator:

- **Use the Crisper Drawer:** Store fruits and vegetables in the crisper drawer of your fridge to maintain optimal freshness and humidity levels.
- **Separate Ethylene Producers:** Keep fruits that produce ethylene, like apples and bananas, away from other produce to prevent premature ripening.
- **Wrapping Greens:** Wrap leafy greens in a dry paper towel and place them in a breathable bag before refrigerating to extend their shelf life.
- **Proper Ventilation:** Ensure your fruits and vegetables have adequate air circulation and are not packed too tightly in the fridge.
- **Temperature Settings:** Keep your refrigerator temperature between 35-40°F (1.6-4.4°C) for the best storage conditions.
- **Freezing:** If you have surplus fruits and vegetables, freeze them at their peak ripeness to preserve their quality for later use.
- **Avoid Washing:** Do not wash produce before refrigerating, as excess moisture can encourage mold growth. Wash them just before eating.
- **Regular Checks:** Regularly inspect your produce for spoilage and remove any items that are past their prime to prevent contamination of other fruits and vegetables.

Temperature Difference (For Storage Rooms Above 32°F (0°C))

The nature of the product plays a crucial role in determining the ideal relative humidity for the storage room. This desired relative humidity then influences the design temperature difference (TD) between the air in the storage room and the refrigerant in the unit cooler. For general storage purposes, particularly for meats, vegetables, and dairy products, it is a standard practice to set the low side of the condensing unit to a TD of 10°F to 12°F. Experience has shown that maintaining this TD allows the cooler to sustain a relative humidity of 80% to 85%, which is optimal for general storage. TDs can be estimated by dividing the unit cooler capacity at a 1° TD by the condensing unit capacity at the desired saturated suction temperature (SST). In low-temperature rooms, the level of dehydration of unwrapped products is directly related to the TD. To prevent excessive dehydration, and since low-temperature condensing unit capacities significantly decrease as the suction temperature drops, it is recommended to limit the maximum TD to 10°F.

Table: 5.2 Recommended Temperature Differences (TD) for Four Classes of Foods (Forced Air Unit Coolers)

Class	T.D. (°F)	Approx. RH	Description of Product Classes
1	7-9 °F	90%	Minimizes moisture evaporation; includes vegetables, flowers, unpacked ice, and chill rooms.
2	10-12 °F	80-85%	General storage, convenience store coolers, packed meats & vegetables, fruits, and similar products. Requires slightly lower relative humidity than Class 1.
3	12-16 °F	65-80%	Includes beer, wine, pharmaceuticals, potatoes, onions, melons, and short-term packaged products. Requires moderate relative humidity.
4	17-22 °F	50-65%	Prep & cutting rooms, beer warehouses, candy or film storage, and loading docks. These require low relative humidity or are unaffected by humidity.

Table: 5.3 Recommended Air Changes per Hour

Type of Application	Recommended Number of Air Changes	
	Minimum	Maximum
Holding Freezer	40	80
Packaged Holding Center	40	80
Cutting Room	20	30
Meat Chill Room	80	120
Boxed Banana Ripening	120	200
Vegetables and Fruit Storage	30	60
Blast Freezer	150	300
Work Areas	20	30
Unpackaged Meat Storage	30	60

Product Safety or Capacity Control

For large boxes, it is advisable to distribute the load across multiple units. If the load exceeds the capacity of a 10 HP unit, splitting it among several units ensures that the customer retains some level of refrigeration in case of mechanical failure.

Additionally, as refrigeration is typically designed to handle the worst 1% of the year's demands, using multiple units offers greater flexibility in managing capacity. In situations where the load is lower, some units can be shut off, and the remaining units will maintain adequate refrigeration with only a fraction of the horsepower needed during peak summer periods. Staggering the startup of multiple units also helps reduce the demand charges imposed by the utility company, leading to lower electricity costs for the customer.

Type of Operation or Air Flow

When selecting and positioning the unit cooler, two key factors must be considered: uniform air distribution and air velocities that are appropriate for the specific application. The airflow direction and the air throw should be designed to ensure air movement is directed towards areas with heat gain, including the walls and ceiling of the room as well as the product itself. It is ideal for the unit cooler(s) to discharge air towards doors or openings, if feasible. Avoid placing the unit cooler near a door, as this may lead to excessive infiltration, which can cause issues like fan icing and hoar-frost.

Additionally, placing a unit cooler in the airflow path of another unit should be avoided, as this can create defrosting challenges. For general storage coolers and holding freezers, there are no specific requirements for air velocities within the room, but the total air supply should be sufficient to achieve 40 to 80 air changes per hour.

APPLICATIONS OF COLD STORAGE

Cold storage is utilized in multiple ways across the food supply chain:

Domestic Refrigeration

- **Refrigerators:** These essential household devices are used to store perishable foods at safe temperatures. Maintaining the correct temperature (around 40°F for refrigerators) is vital for ensuring food safety.
- **Proper Organization:** Thoughtful arrangement of food inside the refrigerator promotes proper airflow and prevents overpacking, which can interfere with efficient temperature regulation.
- **Regular Defrosting:** In manual-defrost refrigerators, regular defrosting is needed to avoid ice accumulation, which can hinder the appliance's efficiency and affect temperature consistency.
- **Freezers:** Freezers provide a long-term storage solution for perishable items at much lower temperatures (around 0°F).
- **Rapid Freezing:** Techniques like flash freezing allow food to retain its quality and texture by reducing the formation of large ice crystals.

Commercial Cold Storage

- **Commercial Cold Storage Facilities:** These expansive storage spaces are designed to meet the needs of food businesses, utilizing state-of-the-art refrigeration systems to maintain precise temperature and humidity conditions for different food products.
- **Controlled atmosphere storage (CAS):** This sophisticated storage method controls the levels of oxygen and carbon dioxide in the environment, helping to prolong the shelf life of specific fruits and vegetables.

Refrigerated Transportation

- **Refrigerated Trucks and Containers:** These specialized vehicles and containers are designed to keep perishable goods at a constant cool temperature during transit, preserving their safety and quality from the producers to distributors and retailers.
- **Temperature Monitoring:** Advanced real-time temperature monitoring systems are commonly used in refrigerated transportation to maintain consistent and optimal temperature control throughout the entire journey.

Ensuring Effective Cold Storage Procedures

To fully harness the benefits of cold storage for food safety, it's vital to follow best practices:

- **Maintaining Optimal Temperatures:** Regularly monitor and maintain the recommended temperature ranges for various food types in refrigerators, freezers, and during transport. Use thermometers to ensure accurate temperature readings.
- **Adhering to FIFO (First-In, First-Out) Method:** Ensure that older items are used first to prevent them from being stored beyond their safe shelf life. Label and date food products for efficient stock rotation.
- **Proper Handling and Packaging:** Practice good hygiene by thoroughly washing hands and using clean utensils. Employ suitable food-grade packaging materials to reduce contamination and preserve food quality.
- **Routine Cleaning and Upkeep:** Regularly clean and sanitize cold storage units to prevent bacterial growth and ensure efficient operation. This includes cleaning shelves, drawers, and defrosting freezers as needed.

By understanding the principles of food spoilage and implementing proper cold storage practices, we can significantly improve food safety, reduce food waste, and ensure the availability of safe and nutritious food for all.

FUNDAMENTALS OF SOLAR COLD ROOM

Solar cold rooms utilize solar panels to generate electricity, which powers the refrigeration system and other essential functions. These systems require consistent maintenance to ensure their efficient and safe operation.

The cold room is designed to maintain temperatures between 4°C and 10°C. It operates without a battery for the compressor, but a small battery is included to support only the auxiliary loads and the control system. Regular upkeep is crucial to maximize performance and longevity.

A cold room is a PUF-insulated space designed to maintain the required temperature and humidity for preserving various commodities. It is equipped with a refrigeration unit that circulates cold air to keep the stored items in optimal conditions.

A solar cold room, in simple terms, is a cold storage facility powered by solar energy for safe storage of a variety of items. It operates using solar photovoltaic power with a battery backup system. This design helps reduce reliance on the grid and facilitates load shifting during peak usage, ensuring efficient and sustainable operation.

Figure: 5.5 Solar cold rooms

Layout of Solar Cold Storage System

The solar cold room is engineered to maintain ideal temperatures for the storage of perishable goods, preserving their freshness and minimizing spoilage. Let's explore in detail how this advanced technology operates, its advantages, and the positive impact it has on communities globally.

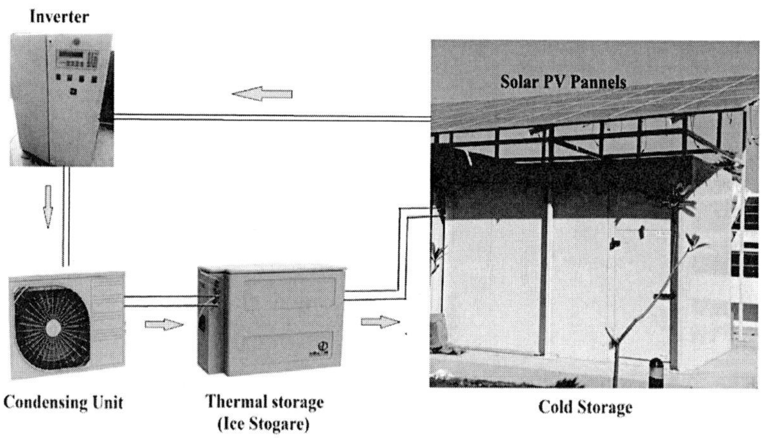

Figure: 5.6 Solar cold storage system layout

Key Features

- **Solar-Powered Efficiency:** Utilizing solar energy for environmentally friendly and sustainable cooling.

- **Energy Independence:** Providing dependable refrigeration without depending on the grid for electricity.
- **Eco-Friendly:** Lowering the carbon footprint and supporting renewable energy solutions.
- **Cost-Effective:** Achieving long-term savings on energy costs and reducing maintenance expenses.

Characteristics of an Excellent Cold Storage

Over time, this commercial machine has become increasingly popular due to the numerous advantages it provides to users. A product's market acceptance hinges on its value to consumers.

- Reduced electricity costs
- Fewer power outages
- Environmentally friendly
- Lower greenhouse gas emissions
- Solar-powered, independent operation
- Minimal maintenance required
- Long-lasting durability

Stand-alone solar power offers an excellent solution for running small cold storage systems in rural areas with limited power load capacity.

Solar Hybrid Cold Room

The adoption of solar hybrid cold rooms is rapidly growing. So, what exactly is this system? A hybrid solar energy system combines both on-grid and off-grid solar energy systems. Essentially, it is a mechanized machine-connected solar power system that includes battery backup, allowing the machine to operate year-round.

Figure: 5.7 Solar hybrid cold room

Is a Hybrid System Worth It?

In today's highly competitive environment, the use of hybrid systems is growing rapidly. This technology helps save money by significantly reducing electricity bills, making the higher initial costs of the system worthwhile. The solar-powered battery has the ability to store most of the energy produced, and you can even sell any excess power generated. For a horticulture farm owner, this system offers remarkable benefits.

Applications of the Machine

Florists and horticulturists often face significant challenges in preserving their flowers and fruits from harsh weather conditions. The solar cold room provides a solution by helping them overcome these challenges and prevent their harvest from spoiling.

Requirement for Cold Storage

After logistics, electricity is the second most significant operating expense in cold storage investments. Bacteria and other microorganisms begin to infest food, causing deterioration, and their activity accelerates as temperatures rise. To keep the temperature as low as possible, these freezer units require continuous power.

Many business owners rely on diesel generators to meet these power needs, but these generators are not only costly but also contribute significantly to pollution. Moreover, with rising power costs and the increasing price of crude oil, diesel becomes more expensive. As a result, store owners face high operating costs, which ultimately impact food prices.

How Cold Storage Powered by Solar Energy Could be the Solution

To understand how solar-powered cold storage can help address this issue and reduce costs for the end user, it's essential to first understand its operation. Solar cold storage systems are typically divided into two types: On-Grid solar-powered cold storage and Off-Grid solar-powered cold storage.

On-grid systems work alongside the electricity grid and do not require energy storage solutions. These systems are commonly used in large-scale cold storage facilities. On the other hand, off-grid storage systems operate independently of the grid and use battery backup to store energy. Smaller and medium-sized cold storage facilities often rely on off-grid solar systems to meet their energy needs.

Types of Solar-Powered Cold Storage

Solar-powered cold storage systems can be classified into three categories based on their size and capacity:

1. Solar-powered Small Cold Storage: These are designed for personal or individual use.

2. Solar-powered Medium Cold Storage: These are intended for small groups or community-level applications.

3. Solar-Powered Large Cold Storage: These are used for large-scale business operations, handling greater volumes of produce.

Advantages of Solar-Powered Cold Storage

Cold storage systems powered by clean energy offer several benefits:

- Decreased reliance on grid supply or fossil fuels
- Lower Greenhouse Gas (GHG) emissions
- Cost-effective operations, reducing electricity expenses
- Improved power reliability
- Helps farmers overcome transportation challenges to access larger storage facilities

Key Points Regarding Solar Panels

- **Energy Generation:** Solar panels produce clean, renewable energy by capturing the sun's power. This energy can be used to power electrical appliances, homes, cooling systems, businesses, or even entire communities, depending on the energy output.
- **Composition:** A solar panel is made up of several solar cells that are interconnected and placed in a frame. The cells are typically made of silicon, but some panels use thin-film cells made from materials like cadmium telluride or copper indium gallium selenide.
- **Efficiency:** Solar panel efficiency measures how much sunlight is converted into usable electricity. Panels with higher efficiency can generate more electricity from the same amount of sunlight. Efficiency typically ranges from 15% to 25%, depending on the technology and manufacturer.
- **Installation:** Solar panels are usually mounted on rooftops or in areas with abundant sunlight. They need to be positioned correctly to maximize sun exposure. Some panels can also be integrated into building materials like solar shingles or solar windows.
- **Grid Connections and Storage:** Solar panels can be connected to the grid, allowing excess electricity to be sent back, creating a grid-tied system. Alternatively, off-grid systems use batteries to store the generated electricity for later use.
- **Environmental Benefits:** Solar panels generate electricity without producing harmful emissions, making them an eco-friendly and sustainable energy source. They help reduce carbon dioxide emissions and dependence on fossil fuels.
- **Maintenance:** Solar panels require minimal maintenance. Regular cleaning to remove dust and occasional inspections to ensure optimal performance are recommended. Typically, they last between 25 to 30 years.

OPTIMAL MAINTENANCE OF SOLAR PANELS

Solar panels require minimal maintenance, with the primary tasks being periodic cleaning and ensuring that no objects obstruct sunlight from reaching them. Since solar panels need direct sunlight to function optimally, any obstructions casting shadows can reduce their efficiency.

These panels are built to endure harsh weather conditions, such as hail, snow, ice, and strong winds, ensuring long-term reliability.

Accumulation of dirt, dust, and debris on the panels' surface can hinder their performance by blocking sunlight. If cleaning is needed, the system should first be turned off. Use lukewarm water from a hose to gently wash the panels. Avoid using soap or cleaning agents, as they may leave streaks or cause damage.

For panels with more stubborn dirt, spray them with lukewarm water and wipe them down with a non-abrasive microfiber cloth. It's best to clean from the roof using a ladder or from a telescopic pole, ensuring safety by wearing a harness.

Cleaning should not be done on a hot sunny day, as quick evaporation of water can leave smudges.

Cleaning Solar Panels Colder During Winter

Snow and ice can occasionally build up on solar panels, but their angled design usually causes it to melt and slide off. If snow or ice remains on the panels, spraying them with lukewarm water can aid in removing it. It is important to avoid using hot water, as the sudden temperature difference can lead to cracks in the panels. Additionally, using a non-abrasive brush to gently clean the panels can help avoid scratching the surface.

How Frequently Do You Need to Maintain Your Solar Panels?

Solar panels typically require maintenance two to four times annually. The dry season and early rainy season are ideal for performing maintenance work in Nigeria, especially after the Harmattan season when dust and debris tend to accumulate. It's also important to watch out for other potential issues like tree sap and bird droppings. Inspecting your solar panels quarterly helps track their performance and identify any problems early on, allowing for quicker resolution.

Additionally, it's wise to check wires and connections to ensure they are functioning properly, as pests can sometimes damage them unnoticed.

How to Know When Your Solar Panels Need Maintenance?

Monitoring the energy output of your solar panels is crucial. A consistent decline in performance often indicates the need for maintenance. While solar panels naturally experience a decrease in efficiency as they age, any significant reduction in performance early on is usually due to dirt or obstruction on the panels. It's important to recognize that the effectiveness of solar panels is influenced by factors such as sunlight availability, geographical location, shading, and the quality and size of the installation. For a more accurate evaluation and optimal design, it's recommended to consult a professional in solar energy.

6

Operation and Maintenance of Ripening Chamber

Environmental factors play a crucial role in determining the quality and shelf-life of fruit during storage. Storage facilities are designed to hold fruit after harvest and before it reaches retail locations, while ripening rooms, often part of these facilities, focus on ripening unripe fruit. Although their processes differ, both storage and ripening facilities aim to control environmental conditions that influence fruit ripening and longevity.

Key factors affecting fruit ripening include temperature, relative humidity, ethylene (C_2H_4), carbon dioxide (CO_2), and oxygen (O_2). These levels vary based on the type of fruit and whether the goal is to store or ripen it. Fruit continues to respire and transpire after being harvested, altering its environment through its physiological changes.

The three primary ways fruit affects its surroundings are:

- **Respiration:** Fruit consumes oxygen to break down sugars and produce heat, water, and carbon dioxide. Lowering the temperature during storage reduces respiration, ripening, and aging. CO_2 and O_2 levels also impact the rate of respiration.
- **Transpiration:** When there is a difference in water vapor pressure between the fruit and the surrounding air, the fruit loses water. This can lead to shriveling, spoilage of quality and flavor, and weight loss. Keeping relative humidity high during storage helps retain moisture and maintain fruit weight, which is important for maintaining its market value.
- **Ripening:** Ethylene, a hormone produced by climacteric fruit, triggers ripening. Climacteric fruits, such as bananas and mangoes, ripen in response to ethylene, while non-climacteric fruits, like strawberries and grapes, can be damaged by ethylene, leading to a shorter shelf life. Controlling ethylene levels during storage is essential to prevent premature deterioration.

UNDERSTANDING FRUIT RIPENING

Fruit ripening is a complex process that involves physiological, biochemical, and molecular changes affecting color, sugar content, acidity, texture, and aroma. It generally makes fruit more edible, palatable, and nutritious.

Fruits are classified as:

- **Climacteric Fruits:** These fruits are harvested when fully mature and continue ripening during storage or transit. Examples include bananas, apples, and avocados.

- **Non-climacteric fruits:** These fruits do not ripen after harvest, so they must be fully ripe when harvested. Examples include citrus fruits, grapes, and strawberries.

Artificial Ripening

Artificial ripening is a controlled technique used to accelerate the ripening process of fruits, aiming to achieve specific characteristics that enhance consumer appeal and boost sales. This method is typically applied to climacteric fruits like mangoes, papayas, and bananas to ensure quicker and more uniform ripening. When conducted with safe and approved ripening agents, artificially ripened fruits are considered safe for human consumption worldwide.

This process also facilitates the transportation of fruits such as mangoes and bananas, which are prone to spoilage and softening after ripening. To prevent loss, these fruits are often shipped while still unripe and are ripened artificially at the destination before being sold.

Provisions for Artificial Ripening of Fruits under Food Safety &Standards Regulations, 2011

The use of acetylene gas, commonly referred to as carbide gas, for artificial ripening of fruits is banned under sub regulation 2.3.5 of the Food Safety and Standards (Prohibition and Restrictions on Sales) Regulation, 2011. However, the regulation permits the use of ethylene gas for artificial ripening, with a concentration of up to 100 ppm (100 μl/L), depending on the type of crop, its variety, and maturity.

Major Issues in Artificial Ripening

Industrial-grade calcium carbide, often referred to as "masala," is illegally used by some dishonest traders to produce acetylene gas for artificial ripening of fruits like mangoes, bananas, and papayas. Calcium carbide contains harmful traces of arsenic and phosphorus, which can cause serious health issues such as dizziness, excessive thirst, irritation, weakness, difficulty swallowing, vomiting, and skin ulcers. The acetylene gas released from calcium carbide is also harmful to handlers.

Furthermore, when the chemical comes into direct contact with fruits, it can leave dangerous residues of arsenic and phosphorus. As a result, the use of this substance for fruit ripening is prohibited in India. Due to the widespread misuse of calcium carbide and the lack of alternative ripening agents, the Food Safety and Standards Authority of India (FSSAI) authorized the use of ethylene gas for fruit ripening in India, as per the notification issued on 23.08.2016.

Utilizing Ethylene Gas (C_2H_4) for Artificial Ripening

Ethylene is a naturally occurring hormone within fruits that regulates the ripening process by initiating and controlling various chemical and biochemical activities. When unripe fruits are exposed to ethylene gas, it triggers the natural ripening process until the fruit begins to produce ethylene on its own in significant quantities.

A Standard Operating Procedure (SOP) has been established to cover all aspects of artificial ripening using ethylene gas. This SOP was created by a Working Group formed by FSSAI, following recommendations from the Scientific Panel on Fruits & Vegetables and their Products (including dried fruits, nuts, salt, spices, and condiments). The Working Group includes members from the Scientific Panel and experts from relevant research institutes.

SOP on Artificial Ripening of Fruits by Ethylene Gas

- **Restrictions:**
  - Any preparation that contains calcium carbide or acetylene gas.
  - Any source of ethylene gas that comes into direct contact with fruits, requiring temperature control.
- **Requirements for Ethylene Ripening System/Chamber:**
  - An airtight room, preferably insulated, for better temperature control.
  - Systems for regulating temperature (cooling and heating).
  - Humidity regulation systems.
  - Proper air circulation and ventilation.
  - Ethylene gas generation or injection system.
  - A reliable power supply (for ripening chambers).
  - A display board showing temperature, relative humidity, ethylene concentration, and CO_2 concentration.

Fruits should be stored in ventilated plastic crates or stackable fruit boxes inside the ripening chamber or temporary structure. Crates should be stacked with a minimum 4-6 inch space from the walls and between adjacent crates. To ensure uniform ripening, airflow should be maintained throughout the chamber. The fruit should occupy no more than 75% of the chamber's or crate's volume during treatment.

Table: 6.1 Requirements of exposure time and ripening temperature for different fruits

Fruit	**Ethylene Exposure Time (hours)**	**Ripening Temperature (°C)***	**Relative Humidity (RH) (%)**
Banana	24-48	15-18	90-95
Mango	24-48	20-22	90-95
Papaya	24-48	20-25	90-95
Pear	24-48	18-22	90-95
Tomato	24-48	18-20	90-95

*Fruits should be transferred to ripening chamber once the ripening temperature is attained after pre-cooling by appropriate method.

CONCEPT OF RIPENING ROOMS

The common practice is to harvest fruits and vegetables while they are still "green" to allow sufficient time for ripening before distribution. The "green" produce ripens naturally, but many times this is accelerated by exposing it to elevated levels of ethylene. However, it is important to ensure the following:

- Ripening rooms should be located in buildings separate from traditional cold storage or Controlled Atmosphere (CA) storage facilities.
- If the ripening room must be located within the storage area, several precautions should be taken:
    - Ethylene should be vented from the ripening room to the outside after the exposure period.
    - Exhaust fans (capable of moving 6 to 8 room volumes per hour) should be used to reduce ethylene levels to around 1%.

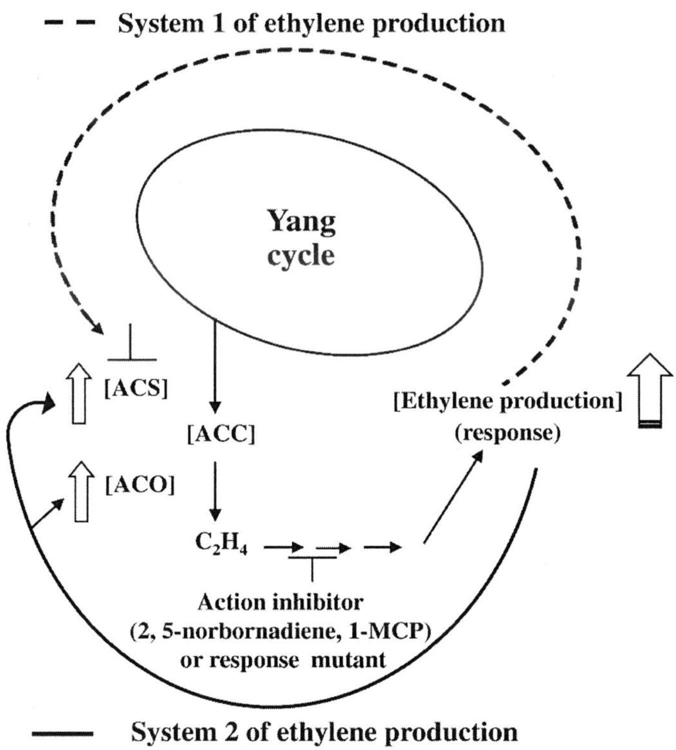

Figure 6.1: Ethylene production

However, due to temperature and humidity concerns, makeup (dilution) air is typically drawn from the refrigerated area surrounding the ripening room. Even after venting, ethylene levels may still remain high enough to continue the ripening process and potentially damage the product. To address this, a "re-circulating air scrubber" or Air & Gas Purification System, with specialized media, should be used to lower the ethylene concentration below harmful levels.

Components of a Banana Ripening Chamber

A typical banana ripening chamber includes the following components:

- Airtight, insulated ripening chamber for maintaining controlled conditions.
- Refrigeration rack system, with an optional standby compressor for reliability.
- Air-cooled or water-cooled condensers to regulate temperature.
- Powder-coated evaporator casing for durability and efficiency.
- Electrical control panel to manage system operations.
- A safety system to protect equipment, including high or low-pressure switches, differential pressure switches, overload protection, and more.
- Ethylene generators or gas-discharge systems specifically designed for ripening processes.
- CO_2 and ethylene analyzers for continuous monitoring of gas levels.

Working Cycle of Banana Ripening Chamber

For bananas, the optimal storage and transport temperature ranges from 13 to 14°C with 90 to 95% relative humidity (RH), while ripening is best achieved at a temperature between 15 to 20°C and 90 to 95% RH. Typically, the banana ripening chamber operates for a 4 to 6-day cycle. The cycle consists of the following steps:

- For ideal operation, conditions should be between 18 to 21°C with RH over 90%, and a pull-down time of 16 to 20 hours to reach 20°C.
- After 24 hours, ethylene gas is injected at a constant room temperature of 18 to 19°C and RH greater than 90%. The ethylene gas quantity is calculated based on the free air space after the bananas are loaded. For example, if bananas occupy 35% of the room, ethylene is calculated based on the remaining 65% free space.
- A holding period of 12 to 16 hours follows, which totals 24 hours after the ethylene injection, with the temperature maintained at 18 to 19°C and RH above 90%.
- After 24 hours, ventilation is initiated to expel ethylene and CO_2, and fresh air is introduced into the cold room. CO_2 levels should not exceed 5000 ppm during ripening.
- The bananas are then held for 3 to 4 more days until the desired yellow color is achieved.
- The temperature is gradually reduced to 14 to 16°C to extend the shelf-life of the fruit.
- An air distribution system ensures uniform ripening by circulating air through palletized crates, equipped with a suitable air bypass sealing system.

These ripening chambers are available in various sizes and refrigeration capacities to meet the needs of traders and farmers.

Figure: 6.2 Fruit ripening chamber

Sources of Ethylene Gas

Ethylene gas for artificial fruit ripening can be obtained through the following systems:

- **For Ripening Chambers**
  - Ethylene gas cylinders
  - Compressed ethylene gas (in aerosol cans)
  - Ethylene generators
  - 2-chloroethylphosphonic acid (Ethephon/Ethrel) combined with alkali
- **For Cartons/Boxes**
  - 2-chloroethylphosphonic acid (Ethephon) in powdered form
  - Ethylene gas encapsulated in biologically safe ingredients such as cellulose, starch, and proteins

Fruit Ripening Under Control in Ripening Chambers

To control the ripening process and ensure the fruit doesn't reach its ideal ripeness before it reaches the supermarket, it's necessary to manage the ripening conditions. This is achieved in ripening chambers where the fruit is placed in transport boxes on pallets or shelves under regulated conditions. The ripening speed can be controlled by adjusting the temperature, humidity, and the targeted supply of ethylene gas, along with managing CO_2 levels.

For instance, bananas typically reach the desired ripeness within 4 to 8 days in these chambers. They require temperatures between 14°C and 23°C (57.2°F and 73.4°F) and a high relative humidity of >90%. To ensure even ripening and avoid CO_2 buildup, it's crucial to maintain consistent air circulation and supply fresh air in the ripening chamber.

Technical Ripening Chamber Equipment

To effectively manage the essential climatic parameters and gas composition in storage, modern ripening chambers are outfitted with the appropriate technical equipment, such as:

- Cooling systems and humidifiers for regulating temperature and humidity.
- Fans and ventilators to ensure proper ventilation and a steady supply of fresh air
- Systems to control the intake and discharge of ethylene, CO_2, and nitrogen.

Additionally, sensors play a critical role by measuring temperature, humidity, CO_2 and oxygen levels, as well as ethylene concentration. These sensors provide the data necessary for optimal control of the ripening process. The accuracy and reliability of these sensors directly impact the success of the ripening process and the quality of the fruit being stored.

- Modern ripening chambers are equipped with a range of technical systems to monitor and regulate the storage environment.

Sensors for Temperature, Humidity, and CO_2 in Ripening Chambers

High humidity presents a significant challenge for the sensors used in ripening chambers. Constant exposure to high humidity can cause sensor drift, leading to inaccurate readings. Additionally, it can result in corrosion of the sensing element and unprotected soldered joints. This not only affects measurement accuracy but also reduces the lifespan of the sensors. Furthermore, since ripening chambers are cleaned between cycles, the sensors can also be exposed to contamination from cleaning agents.

To ensure optimal performance in ripening chambers, sensors must possess the following qualities:

- Long-term stability and high measurement accuracy, even under high humidity conditions
- Resistance to dew formation, dirt, and chemical contamination
- Easy maintenance, such as replaceable sensing probes and filter caps
- Durable enclosures with a high protection rating (IP65 or higher)

Components of a Banana Ripening Chamber

Banana ripening is influenced by numerous factors, and each component plays a crucial role in ensuring the banana reaches its ideal ripening conditions. The following components are essential in the ripening process.

Structure of Insulated Chamber

The most essential element of a banana ripening chamber is its insulated structure. This structure plays a critical role in maintaining consistent temperature and humidity levels, shielding the ripening process from external influences such as weather changes or fluctuations in ambient temperature. These chambers are constructed from high-quality, durable materials, such as polyurethane panels, which ensure effective temperature control and energy efficiency. When evaluating the cost of a banana ripening chamber, it's important to consider the thickness and quality of the insulation. Choosing a chamber with superior insulation will yield better long-term results, making it a more valuable investment.

Refrigeration System

Another component of a ripening chamber is the cooling system, as it helps maintain the correct temperature for ripening. Unlike normal cold storage solutions used to preserve fruit, banana ripening chambers operate at a specific temperature of 13-15°C.

Choosing a room with an efficient cooling system not only optimizes the ripening process, but also reduces operating costs by minimizing energy consumption.

Ethylene Gas Generator

The other component of the banana ripening chamber is ethylene gas. It is a natural plant hormone that promotes and regulates the ripening process of bananas and many other fruits. Speaking specifically about the banana ripening chamber, an ethylene gas generator releases controlled amounts of ethylene to effectively stimulate the ripening process.

Compared to other available alternatives, this is a safer and faster alternative that allows bananas to ripen naturally as well. in large portions. Make sure your chamber releases the right amount of ethylene, as too much gas can accelerate ripening too quickly, while too little can cause inconsistencies.

Humidity Control System

Another important part of the banana ripening room is the humidity control system. The secret to evenly ripening bananas lies in maintaining the proper humidity level, as this directly affects the quality and texture of the banana. One of the main functions of humidity control is to prevent excessive moisture loss, as it will cause the bananas to rot or discolor. For bananas to ripen evenly, the ideal humidity should be around 90-95%, as it keeps the fruit moist and fresh without promoting the growth of mold or bacteria.

To ensure that the banana ripening process proceeds smoothly, the humidity control system includes components such as humidifiers, dehumidifiers, and sensors that continuously monitor and adjust humidity levels.

Ventilation and Air Circulation System

Finally, ventilation and air circulation systems are essential elements of the banana ripening chamber. These systems ensure that other components such as temperature, humidity, and ethylene gas are functioning correctly and are distributed evenly throughout the chamber.

Typically, ventilation and air circulation systems consist of fans, ducts, and air exchange mechanisms, all designed to ensure a continuous flow of fresh, conditioned air, which is crucial for the proper ripening of bananas.

CENTRALIZED ETHYLENE GAS INJECTION SYSTEM FOR RIPENING FRUITS

In a centralized ripening system, fruits are stored in a sealed, insulated ripening room where temperature, relative humidity (RH), and air circulation are carefully controlled and maintained. The fruits are then exposed to a specific amount of Ethylene gas at designated intervals, while the levels of Carbon Dioxide are also monitored and controlled. Air is periodically infused to ensure proper conditions.

Pressurized ethylene gas is gradually released into the ripening room through a double stage gas regulator and solenoid Valve, continuing until the desired concentration is reached and maintained.

Figure: 6.3 Ethylene gas injection system for ripening fruits

Important Features

- Minimal maintenance – saves time and reduces costs
- Complete control over the process

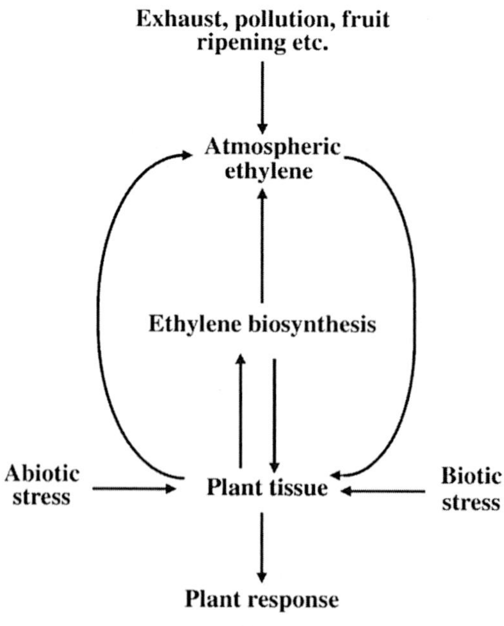

Figure: 6.4 Schematic presentation of ethylene gas in fruit ripening.

- Verifiable ethylene levels for accuracy
- Simple and fast installation process

Figure: 6.5 Banana ripening machine

The Effect of Ethylene

Ethylene has the effect of accelerating ripening, aging, and ultimately causing spoilage in fruits and vegetables. Some of the negative impacts of ethylene on common produce include:

- **Apples:** scalding
- **Eggplant:** browning
- **Bananas:** decay
- **Potatoes:** sprouting
- **Grapes:** mold
- **Onions/Garlic:** odor
- **Broccoli:** yellowing
- **Carrots:** bitterness
- **Carnations:** wilting
- **Green leafy vegetables:** color loss

Even small concentrations of ethylene, from a few parts per billion (ppb) to parts per million (ppm), can reduce the vitality of plants, shorten their lifespan, and degrade product quality. Typically, a concentration of 0.1 ppm is considered the threshold for ethylene's effects on plant growth and produce quality.

Fruits and Vegetables are Affected by Conventional Practices

Closed environments such as truck trailers, shipping containers, warehouses, and cold rooms contribute to the increased concentration of ethylene gas. During the later stages of post-harvest, artificial ripening using ethylene is commonly practiced to ensure produce reaches consumers with the desired ripeness, enhancing its taste, color, texture, and nutritional value. A typical example is the "forced" ripening of bananas during periods of high demand.

However, conventional practices adopted by warehouses and cold stores, such as lowering temperature and humidity and using gases like nitrogen and sulfur, do extend shelf life but come with several drawbacks:

- **Use of Toxic Gases:** The gases involved can be harmful to human health.
- **Health Risks:** Exposure to these gases is dangerous for those handling the produce.
- **Alteration of Natural Fruit Properties:** These methods can change the natural characteristics of the fruits.

While refrigeration and humidity control can slow down decay, they do not stop the production of ethylene gas in storage. Even small amounts of ethylene can accelerate the decay process, making ethylene control essential in the cold chain to maintain the quality and longevity of fresh produce.

Why Ethylene Control?

Advantage of Ethylene Control

- **Prolongs the Shelf Life:** It extends the lifespan of fruits and vegetables after harvest, allowing them to stay fresh longer.
- **Preserves Freshness and Reduces Spoilage:** Warehouse owners can easily maintain the quality of produce, minimizing wastage and spoilage.
- **Meets Rising Demand:** It enables suppliers to cater to the growing market demand for off-season fruits and vegetables, ensuring availability year-round.

Applicationsof Ethylene Control

The three main application areas are as follows:

- **Controlled Atmosphere Storage:** This method involves adjusting the levels of oxygen, carbon dioxide, and humidity to slow down the ripening process and extend the storage life of fruits and vegetables.
- **Fruit/vegetable Ripening Rooms:** These specialized rooms use controlled temperature, humidity, and ethylene gas to accelerate the ripening process, ensuring fruits and vegetables reach their peak quality for sale.
- **Shipping/Transportation of Horticultural Produce**: This process ensures that fresh produce is transported under optimal temperature and humidity conditions, maintaining freshness and reducing spoilage during transit to markets or retailers.

These are described briefly in the following:

Controlled Atmosphere Storage (CA) : The principle of controlled atmosphere (CA) storage for horticultural products is centered around managing two key factors that influence plant aging:

- Reducing respiration rates to slow down the process of aging.
- Delaying ethylene-triggered changes that lead to senescence.

CA storage is achieved through:

- Increasing the concentration of carbon dioxide (CO_2) in the environment.
- Decreasing the amount of oxygen (O_2) available.
- Maintaining low temperatures to slow down biological processes.
- Preventing the accumulation of internally generated ethylene, which can trigger senescence and accelerate the aging process.

By controlling ethylene gas, the quality of horticultural products is preserved, and their shelf life is extended significantly. While refrigeration and humidity control can reduce ripening and decay, they are not sufficient for controlling ethylene, which is a critical factor in maintaining the longevity of the produce.

MAINTENANCE OF RIPENING CHAMBERS: THE CRITICAL ASPECTS

Maintenance of ripening chambers is essential for ensuring the best conditions for fruit ripening, particularly for fruits like bananas, mangoes, and other tropical varieties. Proper upkeep of these chambers helps prevent spoilage, enhances fruit quality, and prolongs shelf life. Below are some critical aspects of maintaining ripening chambers:

Temperature Control

- **Monitor Temperature Consistently:** Regularly check that the chamber's temperature remains stable and within the ideal range for the specific fruit (usually 18–22°C for bananas).
- **Calibrate Temperature Controls:** Periodically calibrate thermostats and sensors to guarantee accurate readings.
- **Maintain Insulation:** Examine the chamber's insulation for any signs of damage. If any issues are found, repair them promptly to prevent heat loss.

Humidity Management

- **Ensure Proper Humidity Levels:** For bananas, the ideal humidity range is 90–95%. Regularly monitor and adjust the humidity using humidifiers or dehumidifiers to maintain the correct levels.
- **Clean and Recalibrate Humidity Sensors:** Keep humidity sensors clean and recalibrate them as needed to ensure accurate measurements.

Ventilation and Air Circulation

- **Maintain Adequate Airflow:** Proper airflow is essential for uniform ripening. Regularly inspect and clean fans and air ducts to prevent blockages and ensure optimal air circulation.
- **Inspect Fans:** Check fans for unusual noise or signs of malfunction. If necessary, replace or repair fans to maintain efficient operation.

Gas Control (Ethylene Gas)

- **Monitor Ethylene Gas Concentration:** Ethylene gas plays a key role in controlled ripening. Regularly monitor its levels to ensure the ripening process is proceeding correctly and the system is functioning as expected.
- **Inspect for Leaks:** Thoroughly check the ethylene gas supply lines and storage tanks for any potential leaks to maintain safety and efficiency.

Cleaning and Sanitization

- **Frequent Cleaning:** Regularly clean the interior of the ripening chamber to prevent mold, bacteria, and pest buildup. Use safe, non-toxic cleaning products to maintain hygiene.
- **Sanitize the Floors:** Thoroughly clean the chamber floors to remove any residue left from previous fruit batches, ensuring a clean environment for the next load.
- **Air Filter Maintenance:** Inspect and clean air filters on a regular basis to prevent blockages, ensuring the ventilation system operates efficiently and the air stays fresh.

Lighting and Sensors

- **Check Lighting Systems:** Ensure that the lighting is functioning properly and does not disrupt the ripening process. Proper lighting is essential to maintain a stable environment for the fruit.
- **Sensor Calibration:** Regularly inspect and calibrate temperature, humidity, and gas sensors to confirm that they are accurately measuring within the desired ranges for optimal ripening conditions.

Safety and Security

- **Safety Inspections:** Regularly check for electrical hazards, verify that fire extinguishers are easily accessible, and ensure emergency protocols are in place and up to date for the safety of the facility.
- **Monitoring System:** Implement remote monitoring systems to continuously track temperature, humidity, and gas levels, ensuring the ripening chamber is operating efficiently and consistently.

Record Keeping and Documentation

- **Track Data:** Keep detailed logs of temperature, humidity, and ethylene gas levels for each batch to track conditions and easily identify any issues that may arise during the ripening process.
- **Maintenance Logs:** Document all maintenance activities, repairs, and inspections to create a reference for future checks and ensure consistency in the ripening chamber's upkeep.

Calibration and Adjustments

- **Routine Calibration:** Consistently calibrate all monitoring and control devices, such as those for temperature, humidity, and gas levels, to maintain accuracy in the ripening process.

- **Adjust Settings as Needed:** Modify the settings of the ripening chambers as necessary based on the specific fruit variety and its ripening requirements, ensuring optimal conditions for each batch.

Schedule for Preventive Maintenance

- **Establish a Maintenance Schedule:** Develop a structured preventive maintenance plan that covers regular cleaning, inspections, and calibrations. This proactive approach minimizes unexpected failures and ensures seamless operation.

By adhering to these maintenance guidelines, ripening chambers will operate efficiently, guaranteeing timely ripening, superior fruit quality, reduced wastage, and extended shelf life.

7

Cold Chain Logistics and Cold Transport

Cold chain logistics refers to the technology and processes that ensure the safe transportation of temperature-sensitive goods throughout the supply chain. It is deeply rooted in scientific principles, assessing the relationship between temperature and perishability to maintain product integrity.

Any item categorized as "perishable" typically requires cold chain management. This includes food products such as meat, seafood, and fresh produce, as well as medical supplies and pharmaceuticals.

Although, modern transportation is a relatively recent development, the practice of moving temperature-sensitive products dates back to the late 1700s when the British used ice to preserve fish. By the late 1800s, ice was widely used to transport perishables. Dairy products were shipped from rural areas to urban markets. Cold chain technology has long been a crucial component of global trade and will continue to play a vital role in the future.

COLD CHAIN LOGISTICS

A cold chain is a specialized supply chain designed for handling perishable and temperature-sensitive goods, commonly referred to as cool cargo. These include fresh produce, meat, dairy, seafood, chemicals, pharmaceuticals, flowers, wine, and other delicate products.

To preserve the quality and integrity of these perishable items, specific low-temperature conditions must be maintained throughout the supply chain. Certain products require refrigeration; others must be kept frozen, while some demand extreme conditions—often referred to as an ultralow chain or deep freeze. Any failure to maintain the appropriate temperature can result in product spoilage, compromising safety and quality, and ultimately leading to significant financial losses.

Cold chain logistics encompasses a series of coordinated activities designed to safely handle, store, and transport temperature-sensitive products from the manufacturer or supplier to the end consumer. This process ensures that perishable goods maintain their required temperature conditions throughout the supply chain, preserving their quality, safety, and efficacy.

The longer the distance between the original supplier and the end consumer, and the more stages or transportation modes involved, the harder it becomes to maintain the required temperature. In international shipping, for instance, freight often passes through multiple storage facilities or distribution centers and undergoes several reloading processes. Unless the goods are kept in a sealed refrigerated container at all times, these transitions increase the risk of temperature fluctuations.

Within cold chain logistics, the most challenging stages include packaging, storage, and shipping.

Cold chain logistics is a complex process with multiple components, including:

- **Cold Storage:** Facilities that hold perishable goods until they are ready for transport.
- **Cooling Systems:** Equipment that ensures products are maintained at optimal temperatures during processing, storage, and transportation.
- **Cold Transport:** Vehicles and systems that maintain stable temperature and humidity levels during transit.
- **Cold Processing:** Facilities that allow for the safe handling and processing of goods while maintaining hygiene and sanitation.
- **Cold Distribution:** The organized loading and transportation of crates, boxes, or pallets to distribute perishable products efficiently.

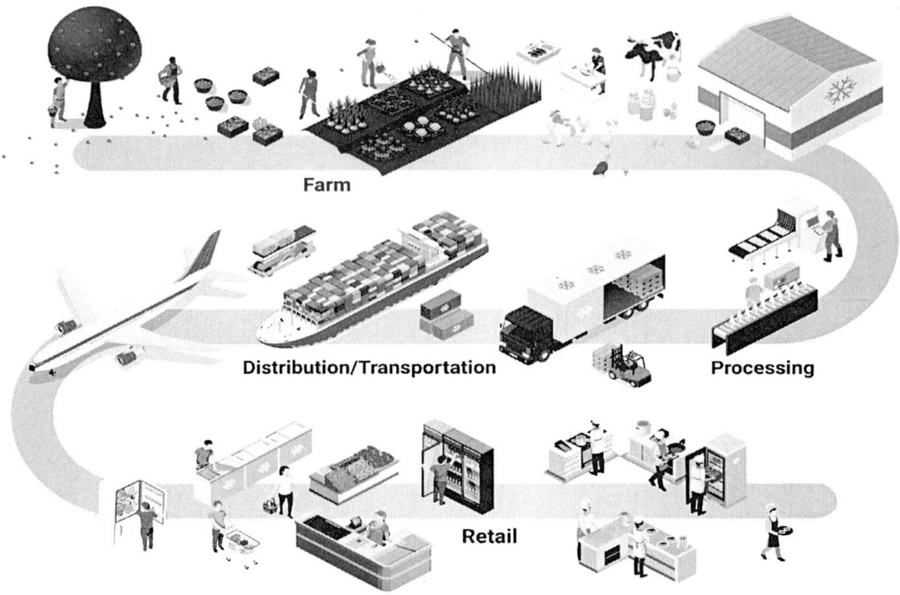

Figure: 7.1 Processes involved in cold chain logistics

Cold chain logistics involves the careful handling, storage, and transportation of temperature-sensitive goods—such as pharmaceuticals, food items, and chemicals—within a strictly controlled temperature range. This ensures product integrity is maintained from the point of origin to the final destination.

Storage in the cold chain relies on specialized facilities designed to keep goods at precise temperatures. These include the following:

- **Refrigerated Warehouses**: Large-scale storage facilities equipped with advanced refrigeration systems to maintain stable product temperatures.
- **Cold Rooms:** Smaller, temperature-controlled spaces used for storing perishable goods. Larger cold storage facilities often contain multiple cold rooms to accommodate various storage needs.

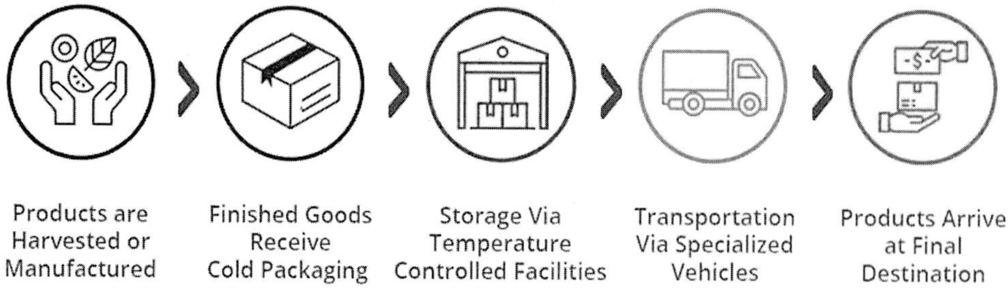

Figure: 7.2 Cold chain logistics process

Impacts of Cold Chain Logistics

- **Integrated Cold Chain Networks:** Cold storage plays a vital role within the larger cold chain logistics system, which encompasses transportation and distribution. Recognizing the interdependence between cold storage and the entire supply chain is essential for optimizing logistics, maintaining seamless connectivity, and reducing disruptions.
- **Data-Driven Decision-Making:** Leveraging data analytics in cold chain logistics improves decision-making by offering valuable insights into demand forecasting, inventory management, and route optimization. This data-centric approach enhances the overall efficiency, reliability, and effectiveness of cold storage operations.

Regulatory Compliance and Global Standards

- **Aligning Global Regulations:** As the food and beverage industry continues to expand internationally, aligning regulations and standards for cold storage is essential. Standardizing safety protocols ensures consistency, facilitates smoother cross-border trade, and strengthens global food security.
- **Commitment to Ongoing Enhancement:** Cold storage facilities must proactively adapt to changing regulations. Conducting regular audits, compliance checks, and continuous improvement initiatives helps uphold the highest standards of food safety and storage efficiency.

Ultimately, cold storage serves a far greater purpose than just preservation. It integrates sustainability, technological innovation, logistics optimization, workforce development, and regulatory adherence. By embracing these elements, the industry can enhance the effectiveness of cold storage while contributing to a more resilient and sustainable global food supply chain.

TECHNOLOGIES FOR THE COLD CHAIN

Cold chain transportation utilizes various methods to maintain goods at the required temperatures. Factors such as transport duration, shipment size, and seasonal conditions determine the most suitable approach.

- **Gel Packs:** Commonly used for medical and pharmaceutical shipments to maintain controlled temperatures.
- **Dry Ice:** Ideal for keeping goods frozen for extended periods; frequently used for pharmaceuticals, food, and hazardous materials. Unlike regular ice, it sublimates instead of melting upon air exposure.
- **Liquid Nitrogen:** Primarily used for transporting biological cargo, including organs and tissues. With extremely low temperatures, it ensures prolonged freezing.
- **Eutectic Plates:** Also known as cold plates, these function similarly to gel packs but offer the advantage of being reusable.
- **Reefers:** Insulated, temperature-controlled vans, semi-trucks, or ISO containers designed to regulate airflow and maintain stable temperatures.
- **Quilts:** Insulated coverings placed over freight to help retain consistent temperatures, ensuring frozen items remain frozen for extended durations.

COLD CHAIN STANDARDS & REGULATIONS

Regulatory Framework and Compliance in the Cold Chain Industry

The cold chain industry operates under an intricate network of regulations, which differ based on regional requirements and the type of product being handled. Adherence to these regulations is crucial for ensuring the safety, quality, and efficiency of perishable goods throughout the supply chain. Several key regulatory bodies oversee compliance and establish guidelines for industry operations.

Major Regulatory Authorities

Major regulatory authorities are as follows:

1. **Food Safety and Standards Authority of India (FSSAI):** Governs food safety norms and ensures adherence to quality standards in India's cold chain sector.
2. **Central Drugs Standard Control Organization (CDSCO):** Supervises and regulates the storage and distribution of pharmaceuticals, ensuring compliance with safety guidelines.
3. **Ministry of Agriculture & Farmers Welfare:** Manages regulations related to agricultural commodities, overseeing cold storage infrastructure and facilities.
4. **Bureau of Indian Standards (BIS):** Establishes benchmarks and technical standards for various products, including refrigeration and cold storage equipment.

Core Compliance Areas

Core compliance areas for cold chain are as follows:

1. **Temperature Regulation:** Maintaining precise temperature levels for perishable and temperature-sensitive goods is of utmost importance. Indicold integrates advanced monitoring technology to ensure that temperature conditions remain stable during storage and transportation, thus aligning with the FSSAI and CDSCO guidelines.
2. **Documentation and Traceability:** Keeping comprehensive records is fundamental to ensuring accountability and traceability in cold chain operations. Our real-time tracking systems facilitate the seamless monitoring of products, providing complete documentation to verify compliance with FSSAI regulations and ensuring product integrity throughout the supply chain.
3. **Facility Compliance and Standards:** Cold storage facilities must adhere to rigorous regulatory standards, covering aspects such as insulation, hygiene, pest control, and overall facility management. Frequent audits and inspections help maintain compliance with the BIS standards and Ministry of Agriculture & Farmers Welfare guidelines, ensuring that all facilities meet prescribed industry norms.
4. **Employee Training and Standard Operating Procedures (SOPs):** A well-trained workforce is essential for maintaining regulatory compliance and upholding industry best practices. We prioritize ongoing employee training programs, ensuring that all staff members remain proficient in FSSAI and CDSCO regulations and adhere strictly to defined Standard Operating Procedures (SOPs) for cold chain management.

By strictly adhering to these regulatory mandates, the cold chain industry ensures product safety, operational efficiency, and legal compliance, fostering reliability and trust across the supply chain.

Compliance with Regulations

Regulatory compliance plays a critical role in cold chain logistics, as it ensures the maintenance of product quality, prevents spoilage, and upholds customer safety. By adhering to industry regulations, cold chain operations can safeguard temperature-sensitive goods, meet international and local standards, and ensure the integrity of products throughout their journey from origin to destination.

Storage & Transport Temperature Ranges for Various Products

- **Fruits:** 0°C to 5°C – Slows down ripening, prevents spoilage, and keeps fruits fresh.
- **Dairy Products:** 1°C to 3°C – Ensures proper preservation and quality.
- **Pharmaceuticals:** 2°C to 8°C – Essential for maintaining the potency and effectiveness of medicines and vaccines.
- **Frozen Food:** Below -18°C – Prevents bacterial growth and preserves food in a frozen state.
- **Seafood:** 0°C – Prevents spoilage and maintains quality.

Procedure of Cold Chain Logistics Management

- **The Preparation Stage:** Cargo refrigerators are primarily designed to maintain temperature rather than actively reduce it. Therefore, before transporting a temperature-sensitive product, it must first be cooled down to its optimal storage temperature.
- **Selecting the Right Path:** When choosing the best route for shipping, it's not just about selecting the shortest path. Factors such as road conditions, traffic likelihood, and weather conditions should also be considered to ensure safe and timely delivery.
- **Customizing Procedures:** Tailoring the logistics and delivery process to suit specific needs can significantly improve efficiency. For instance, for large quantities of cold chain products, refrigerated trucks are the best choice. However, for smaller shipments with urgent delivery needs, a van may be a more suitable option.
- **Adjustment for Last Mile:** The "last mile" refers to the final stage of the delivery process. This part of the journey is often considered the most critical, as any issues during this stage, such as spoilage, could render all prior efforts in vain.
- **Quality Check Assurance:** Upon receiving the shipment, the recipient should verify the cold chain conditions by checking for any temperature deviations. This process is vital for maintaining trust and accountability throughout the supply chain.

Common Cold Chain Management Issues

Cold supply chain management faces several challenges that can significantly impact freight shipments. Drivers are often very aware of these potential issues and take proactive steps to prevent problems before they arise.

Common challenges include the following:

- **Product Quality Issues:** Maintaining the quality of perishable items, such as food and produce, starts with proper sanitizing, cleaning, and sorting before packaging and loading.
- **Inadequate Packaging:** Proper packaging is crucial to prevent contamination and damage during transit. Airflow and ventilation must also be considered to avoid degradation.
- **Lack of Proper Documentation:** Comprehensive documentation is essential throughout the cold supply chain, especially during transit. Data loggers that track temperature and environmental conditions are vital in ensuring the load remains in optimal conditions.
- **Shipment/Transport Delays:** Delays can severely affect the integrity of cold chain shipments, as time-sensitive conditions, like maintaining the correct temperature, are critical for perishable goods.
- **Disrupted Climate Control/Temperature Fluctuations:** Variations in temperature can occur due to factors like frequent door openings, improper

pre-cooling, extreme weather conditions, or malfunctioning cooling systems. These fluctuations can compromise the quality of the goods.

COLD TRANSPORT

Cold transport involves the transportation of temperature-sensitive goods while maintaining the desired temperature throughout the journey. This is achieved through the use of various temperature-controlled vehicles and containers:

- Refrigerated trucks and containers are equipped with built-in refrigeration units, which may use liquid nitrogen, to sustain constant temperatures during long-distance transport.
- Refrigerated ships are designed with large, refrigerated cargo holds to transport goods by sea while ensuring proper temperature control.
- Airplanes with temperature-controlled cargo holds provide fast transport over long distances, ensuring that goods remain at the correct temperature during transit.

Figure: 7.3 Cold transport logistic

Cold chain logistics involves various packaging components designed to maintain the safety and effectiveness of temperature-sensitive products throughout the supply chain.

COLD CHAIN STORAGE

Warehousing is a multifaceted system that includes various processes such as storage, consolidation, deconsolidation, and loading/unloading. In cold chains, specialized warehouses often provide additional services like blast freezing, precooling, frozen receiving, ripening, and inspection checks.

It is crucial at this stage to design suitable facilities with effective climate control for both storage and loading areas, and to implement best practices when handling perishable goods.

Cold Chain Elements

A cold chain can be understood as the close integration of three key technologies.

1. **Product**: Each product has specific physical characteristics that require precise temperature and humidity conditions. These conditions determine its transportation method, ensuring the product's attributes remain intact

throughout the journey. If not properly managed, the product may lose its value either partially or entirely, depending on its perishability and fragility.

2. **Origin/Destination:** This refers to the locations where temperature-sensitive products are produced and consumed. It highlights the challenges in making products available in markets far from their production sites, often due to logistical constraints. However, advances in cold chain logistics have enabled sourcing strategies from increasingly distant locations, often spanning the globe.

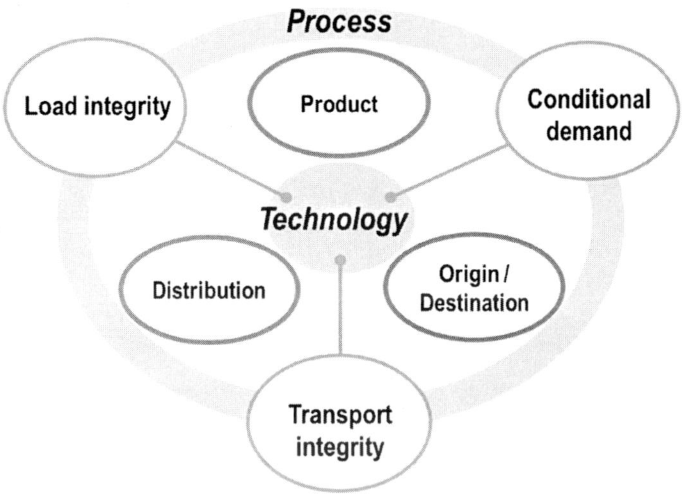

Figure: 7.4 Elements of the cold chain

3. **Distribution:** This involves the available methods and infrastructure for transporting a product in a temperature-controlled environment. It includes refrigerated containers (reefers), temperature-controlled trucks, and specialized warehousing facilities.

Chilled Transport Van

A chilled transport van, sometimes referred to as a reefer van or refrigerated van, is a vehicle that can carry perishable goods at a constant temperature. The food, pharmaceutical, and floral industries are just a few of the industries that use these vans.

Semi Chilled Transport

Semi-refrigerated transport, also known as refrigerated transport, is the process of transporting perishable goods at a low and controlled temperature. This is done using specially designed trucks and trailers with integrated refrigeration systems.

Semi-refrigerated vehicles are another transport system used for perishable goods. They are usually equipped with a refrigeration and insulation unit to maintain temperatures down to -20 °C (-4 °F). The thermal reading can be adjusted depending on the type of goods being transported.

Refrigerated Van System

A refrigerated vehicle, sometimes referred to as a refrigerated van or truck, is a type of insulated transport vehicle used to maintain a consistent temperature for goods. Cooling technologies like a refrigeration system, thermoelectric cooling, or evaporative cooling are found in almost all refrigerated units. This keeps things cold during transit and aids in regulating the temperature inside the car.

Types of Refrigerated Vehicles

Let's explore the various types of refrigerated trucks and vans available on the market today:

- **Insulation Vans:** These are standard vehicles retrofitted with insulation to slow down the transfer of heat, helping to keep the interior cooler for longer periods. While they don't come with a built-in refrigeration unit, ice or cold packs can be added to maintain lower temperatures for specific periods. The interior is usually equipped with 50mm Styrofoam or polystyrene insulation, which prevents external heat from entering. These vans are perfect for short trips and are a cost-effective, energy-efficient option for transporting less perishable goods.
- **Chiller Conversion Vans:** Chiller vans are similar to insulation vans but come equipped with refrigeration systems that introduce cool air to maintain the freshness of perishable items that don't require freezing temperatures. These vans are essential for goods such as flowers and beverages, keeping them in a cool environment throughout transport. They're indispensable for supply chains where products must stay cool but not frozen.
- **Semi-Freezer Vans:** Semi-freezer vans offer superior insulation and higher refrigeration capacity compared to chiller and insulation vans. They maintain consistently low temperatures, which can be adjusted depending on the products being transported. These vans feature quick defrost mechanisms that prevent frost accumulation and ensure optimal operation of the refrigeration system. Semi-freezer vans combine advanced technology with reliable performance, making them an essential tool for cold supply chains.
- **Full-Freezer Vans:** Full-freezer vans are the top choice for refrigerated transport, equipped with robust freezing capabilities for precise temperature control. These vehicles are ideal for shipping goods that must remain frozen during transit, such as meat, seafood, and dairy products. Their thicker, more heat-resistant doors help protect the cargo from external heat. Full-freezer vans are crucial for maintaining the integrity of highly perishable items, ensuring they meet strict food safety standards.

FACTORS TO CONSIDER IN SELECTION OF A REFRIGERATED VEHICLE

For efficient supply chain management, particularly for perishable goods, refrigerated vehicles must maintain cool temperatures. Many businesses equip trucks, ships, and even train cars with refrigeration units to safely transport their products.

When selecting the right refrigerated vehicle for your operations, it can be challenging due to the variety of options available. To simplify the process, consider the following essential factors:

- **Load Size:** It's crucial to evaluate the size of your cargo before choosing a refrigerated vehicle. Consider how much space your goods will occupy and select a truck or vehicle that can comfortably accommodate the load.
- **Temperature Requirements:** Different goods require different temperatures. For items that only need cool conditions (0-5°C), chiller vans are the ideal choice. For products that need sub-zero temperatures (-18 to 0°C), full-freezer vans are the appropriate option to maintain the required temperature.
- **Location of Delivery:** The delivery location plays a key role in selecting the right vehicle. For long-distance or large-area deliveries, insulated or semi-freezer vans with better insulation and refrigeration control might be more cost-effective compared to smaller options.
- **Budget:** Cost is always a significant factor. Refrigerated vehicles come in various price ranges depending on their model and features. However, investing in high-quality refrigerated vehicles is crucial to ensuring safe and timely delivery of your goods.

Cold Chain Packaging

To maintain the ideal temperature for perishable goods during transportation, it's essential to choose the right packaging, containers, and refrigeration methods. The selection process depends on several factors, including:

- Type of cargo
- Required temperature and humidity levels
- Size of the shipment
- Duration of transit
- Outside temperature (especially important for long-distance shipments or those affected by climate changes)

Different types of temperature-sensitive shipments require specific packaging. For instance, vaccines are transported in small, insulated boxes to ensure proper cooling, while food items are typically transported in 53-foot refrigerated trailers (reefers) equipped with built-in refrigeration units. Packaging technologies will be further discussed in the next section.

Cold Chain Shipping and Tracking

Transporting temperature-sensitive goods, also known as cool cargo, requires specialized refrigerated equipment, such as reefers or freezers. These are custom trucks, railcars, cargo ships, and aircraft designed to maintain the required temperatures. In many cases, it's not just the temperature that needs to be controlled, but also the humidity, so the equipment must be capable of regulating both.

Additionally, while tracking shipments is important in any supply chain, monitoring the environmental conditions of cold chain cargo is critical. It ensures the efficiency of the process and helps maintain product integrity. Shippers, carriers, and consignees all need access to environmental monitoring data for quality assurance and performance evaluation throughout the journey.

Cold Chain Logistics use Cases

Cold chain logistics is essential across various industries, including agriculture, manufacturing, and healthcare, as it ensures the safe storage and transport of perishable goods. The demand for cold chain services continues to grow due to factors like evolving consumer preferences, the "farm to fork" movement, and the increasing need for new drugs and vaccines. Let's take a deeper look at the sectors that rely most heavily on cold chain logistics.

COLD CHAIN IN FOOD INDUSTRY

Storing and transporting perishable food products is the most established and long-standing application of cold chain logistics, which dates back to prevent spoilage and mold. We've learned both in school and through personal experience that the quality of certain foods deteriorates over time due to natural chemical reactions, which can be slowed by maintaining lower temperatures.

Different types of food require varying thermal conditions are given below:

- **Tropical Fruits:** Products like bananas and pineapples require controlled ripening during transport, needing a stable temperature between 12° to 14°C (53° to 57°F).
- **Chilled Goods:** Most vegetables, fruits, dairy products, and meats need refrigeration at temperatures around 2° to 4°C (35° to 39°F).
- **Frozen Goods:** Products such as meats, bread, and cakes require freezing at temperatures between -10° to -20°C (14° to -4°F).
- **Deep-Frozen Items:** Seafood, ice cream, and similar products must be kept deep-frozen, usually between -25° to -30°C (-13° to -22°F).

Temperature Tracking in Transportation and Storage

Real-time tracking of product temperature, humidity, and condition across the supply chain is made possible by temperature monitoring technologies like RFID tags and Internet of Things sensors. If issues occur, this data enables prompt remedial action, guaranteeing that the products continue to be high-quality and safe. In cold logistics, it is critical to continuously monitor the temperature.

Cold Chain Equipment and Packaging Options

As mentioned earlier, reefer trailers, various containers, and specialized vessels are commonly used to transport temperature-sensitive products. Liquefied gases, such as oxygen, nitrogen, argon, and CO_2, are transported and stored in specialized cryogenic tanks, trailers, and railcars.

For storage, warehouse and terminal facilities, as well as distribution centers, are equipped with large walk-in coolers and refrigerators that maintain a constant temperature. For storing pharmaceuticals, drugs, tissue samples, and other medical items, a wide variety of laboratory refrigerators and freezers are available to meet the required temperature specifications.

When it comes to packaging, there are a broad range of cooling products designed to keep perishables at the right temperature. These cooling technologies fall into two main categories:

- **Passive Cold Containers:** These rely on materials such as dry ice, gel packs, gel bottles, liquid nitrogen, eutectic plates, quilts, and other cooling methods to keep the internal temperature low. For pharmaceutical products, which may require ultra-low freezing, specialized containers are being developed to meet the specific needs of the industry.
- **Active Cool Containers:** These containers are equipped with powered refrigeration units that can monitor and control the internal temperature. They come with sensors, a control unit, and ventilators that automatically adjust the temperature to maintain stability.

While active containers may initially seem more expensive, they often provide more cost savings compared to traditional passive ones. These containers are reusable, take up less space, require minimal handling, and eliminate the need for replacing cooling materials. They also offer more reliable temperature control and can be shipped alongside other cargo or as Less Than Truckload (LTL) freight, which is more economical than using dedicated reefers.

Additionally, active containers are equipped with monitoring devices, enabling real-time tracking of shipments and associated activities. Let's explore the technologies that make this possible.

DIGITAL COLD CHAIN TECHNOLOGY

As highlighted earlier, complete visibility in cold chain logistics is crucial. Many temperature-sensitive products are also time-sensitive, and even when stored properly, they need to be transported promptly without any delays.

Therefore, all stakeholders involved must track this vulnerable cargo and continuously monitor its condition. This ensures that everyone stays informed about the status of the goods and can react quickly if any issues arise. Modern digital technology plays a key role in enabling this real-time tracking and monitoring, ensuring the cargo stays within the desired conditions throughout its journey.

Tracking of Cold Chain Lagistics

For tracking cold chain logistics, various technologies can be leveraged to ensure proper monitoring and location tracking:

- Global Positioning System (GPS)
- Global System for Mobile Communication (GSM)
- Radio-Frequency Identification (RFID)
- Bluetooth Low Energy (BLE)

When cold cargo is transported by truck, telematics solutions powered by GSM/GPS enable real-time tracking of the vehicle's location. This location data is typically sent to the carrier's Transportation Management System (TMS) or other related business software. Many carriers offer customers access to tracking information through portals or by providing a link to a third-party tracking service provider.

For further tracking, battery-powered GPS devices can be attached to shipping containers, allowing their location to be monitored across trucks, rail, or ships. Some GPS trackers come with additional monitoring features, providing more comprehensive data for cold chain management.

In maritime shipping, major ocean carriers share their vessels' satellite location data via websites, apps, or customer portals, and third-party logistics (3PL) providers and freight forwarders also offer tracking services. Independent global sea data providers like Marine Traffic provide real-time vessel positions, giving easy access to cargo tracking around the globe. Technologies like RFID and BLE operate differently. RFID tags or beacons need to be attached to containers or individual packages to gather data through scanning. RFID tags can be passive (activated by the reader) or active (with an inbuilt battery that broadcasts signals and can be scanned from a greater distance). BLE beacons work similarly and can be scanned with smartphones, making them a cost-effective solution.

Readers for RFID and BLE are installed at checkpoints like terminals or distribution centers. These systems automatically scan all the tags or beacons within their range. Handheld RFID readers are also available, while BLE can be scanned easily using smartphones.

GPS devices are ideal for real-time, global container or equipment tracking (though they are expensive with shorter battery life), while RFID and BLE technologies are best suited for indoor use or periodic information tracking of individual packages.

Monitoring

The GPS and RFID technologies discussed earlier are great for location tracking, but they don't offer the ability to measure, store, and transmit environmental data. To monitor factors like temperature, humidity, and vibration in cold chain logistics, additional devices are used to ensure the safety and stability of the products. Below are some common options:

- **Data Loggers:** These are the most common monitoring tools. Data loggers are placed inside containers or storage areas to measure temperature (and sometimes humidity) at regular intervals. They record shipping and storage conditions, and data can be retrieved by physically connecting to the device. While cost-effective, their major drawback is that they don't provide real-time monitoring and require close proximity to collect the data.
- **Bluetooth Temperature Sensors (BLE Technology):** More advanced than data loggers, Bluetooth sensors can not only track temperature but also add location data to the measurements. These devices can transmit records wirelessly when scanned with a reader or smartphone. However, they don't offer real-time updates by themselves. To enable live monitoring, these sensors need to be connected to an IoT infrastructure.

- **Internet of Things (IoT):** IoT enables real-time data collection and monitoring, which is essential for cold chain logistics. In this setup, several components work together:
  - **BLE Sensors**: These sensors are attached to containers, pallets, or packages to measure temperature, humidity, CO_2 levels, and other factors.
  - **Portable Gateway Hotspot:** This device collects measurements from the BLE sensors and transmits the data to the cloud.
  - **Cloud Storage:** The cloud stores, processes, and allows for easy retrieval of the data.
  - **User-facing Apps or Platforms:** These apps, such as ERP, TMS, WMS, or other business solutions, display the data through dashboards and reports, making it accessible for users.

IoT-based monitoring offers near real-time updates and greater insight into cold chain conditions, allowing stakeholders to respond to any deviations promptly.

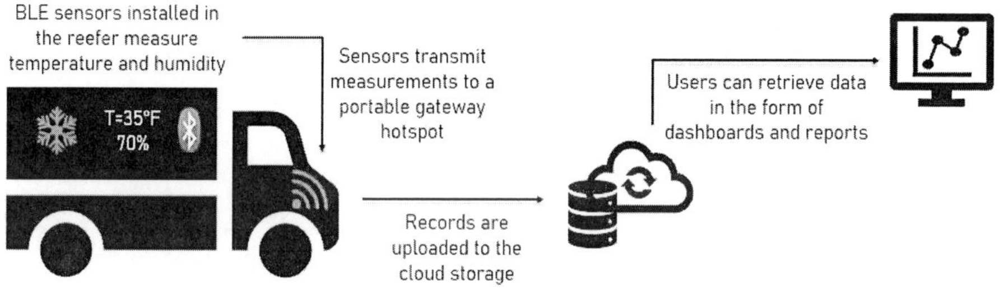

Figure: 7.5 Working of IoT in cold chains

Typical IoT Architecture in Cold Chains

Smart systems in cold chain logistics go beyond real-time monitoring; they can also be set up to send alerts in case of temperature deviations or other abnormal events. These systems help ensure that any critical changes in conditions are promptly addressed, reducing the risk of spoilage or damage to temperature-sensitive goods. By providing immediate notifications, stakeholders can take quick action to maintain optimal conditions and prevent costly losses.

Integrating Tracking and Monitoring Solutions

If you're dealing with just a few shipments in your cold chain, you can easily track them through a customer portal provided by the carrier. However, if your business involves regular shipments, you'll need a more efficient way to monitor your cold cargo and ensure its safety.

While you can opt for GPS tracking devices or set up an IoT infrastructure to track critical indicators, if you prefer the carrier to handle it, you might consider system integration. This option requires development, time, and investment but offers the benefit of convenience and time savings.

By establishing an API connection, you can receive real-time data on location, temperature, and other key metrics directly into your internal systems. This ensures you can monitor your shipments effectively and take immediate action if any disruptions occur.

Cold Chain Logistics Challenges

Shippers are willing to invest in secure cold storage and transportation to extend product shelf life and meet customer demand. However, carriers and 3PL companies face several challenges that make managing cold chains a complex business.

- **Infrastructure Development and Maintenance:** According to the 26th Annual Third-Party Logistics Study (released in September 2021), the majority of 3PLs identified creating and maintaining cold chain infrastructure and temperature monitoring as their main concerns. To fulfill shippers' demands, 3PLs need a variety of equipment, ranging from coolers to deep-freezers, and must continually invest in new technologies to stay competitive.
- **Proper Handling:** The same study revealed that proper handling of temperature-sensitive goods was the biggest challenge for shippers. Ensuring product quality requires strict adherence to practices like sanitation, packaging, cleaning, and sorting.
- **Regulatory Compliance:** With a variety of regulations governing the industry, supply chain participants must stay vigilant and invest in modernization to meet compliance standards. This includes ensuring goods are traceable and visible throughout the supply chain, as well as maintaining accurate temperature records.

Other challenges in the industry include the shortage of skilled specialists and high electricity costs. Despite these issues, 91% of shippers, along with all the 3PLs surveyed, expect an increase in demand for cold chain capacity over the next three years. Smart businesses recognize the need for modernization and investment, as the future of cold chains lies in innovative technologies.

8

Cooling Systems /Cold Chain Technology

Cooling systems play a vital role in cold chain logistics, serving as essential components that regulate temperature for storage, processing, and transportation. An efficient cooling system ensures that food and other perishable products remain fresh for extended periods, even before they begin their journey. Various cold chain products require specific temperature settings to maintain their quality and integrity until they reach their final destination.

For instance, tropical fruits such as bananas and pineapples must be stored at a consistent temperature between 12°C and 14°C. If the temperature exceeds this range, the fruits may ripen prematurely, compromising their quality. Likewise, food items like milk, vegetables, and other perishable fruits should be cooled within the range of 2°C to 4°C. This temperature effectively eliminates harmful pathogens and prevents bacterial growth, preserving the freshness of the produce.

For businesses dealing with frozen food, seafood, or ice cream, a more advanced cooling system is necessary—one that can maintain temperatures as low as -30°C. Such systems are essential to prevent thawing and spoilage, ensuring that frozen goods remain in optimal condition throughout their transportation and storage.

Maintaining cold conditions during transportation is crucial to ensuring the quality and safety of temperature-sensitive goods. Proper temperature control helps prevent spoilage, maintains product integrity, and ensures compliance with industry standards. Cold chain regulations are designed to safeguard public health, uphold the effectiveness of pharmaceuticals, and preserve the nutritional value and flavor of food products.

These regulations establish strict guidelines for handling, storing, and transporting perishable goods, ensuring they remain within the required temperature range throughout the supply chain. Adhering to these standards is essential for preventing contamination, reducing waste, and delivering safe, high-quality products to consumers.

THE TECHNOLOGY OF THE COLD CHAIN

Various technologies work together in a coordinated sequence to maintain the integrity of the cold chain:

- **Monitoring:** Advanced monitoring systems track temperature, humidity, and other critical conditions throughout all stages of the cold chain, from refrigerated transport (reefers) to cold storage warehouses. These technologies help ensure product integrity and detect potential vulnerabilities. The ISO 10368 standard, established in 1992, provides guidelines for reefer temperature monitoring. Older methods, such as Partlow recorders that manually logged temperature fluctuations on a rotating disk, have been replaced by modern electronic devices capable of remote monitoring and real-time data communication.

- **Fabrication:** Cold chain products, such as food and pharmaceuticals, are manufactured and processed in specialized facilities using advanced equipment. For example, blast freezers rapidly freeze meat, preventing the formation of large ice crystals that could degrade quality. To preserve temperature integrity during transit, various packaging solutions, including perforated boxes, crates, and vacuum-sealed packaging, are employed to enhance protection and extend shelf life.
- **Storage:** Cold chain goods are typically stored in refrigerated warehouses or specialized distribution centers before reaching their final destination. These facilities are designed to maintain specific temperature ranges while optimizing energy efficiency. Large-scale refrigerated storage ensures that perishable goods remain fresh and safe until they are ready for transport or sale.
- **Terminal Handling:** With the increasing international trade of perishable goods, transport hubs such as ports and airports have dedicated areas for cold chain logistics. Many container port terminals feature designated spaces for storing refrigerated containers, while some include on-dock cold storage warehouses for improved handling efficiency.
- **Transport:** The transportation of cold chain goods relies on advanced refrigerated technologies. Reefer vehicles, including refrigerated trucks, maritime containers, and unit load devices, are commonly used for temperature-controlled transportation. These systems often depend on built-in refrigeration units powered by generators to maintain stable conditions throughout transit.

Each of these technologies plays a crucial role in ensuring the safe and efficient movement of temperature-sensitive products within the cold chain.

TECHNOLOGIES FOR COLD CHAIN STORAGE

In the intricate world of global logistics, the cold chain represents a highly sensitive and essential operation. Maintaining precise temperature conditions for perishable goods, whether food or pharmaceuticals, is critical. Below is a compilation of cutting-edge cold chain storage technologies designed to regulate and monitor temperatures effectively:

- **Advanced Refrigeration Systems:** Modern refrigeration systems in large-scale warehouses are far more sophisticated than earlier cooling methods. They employ variable frequency drives that adjust cooling intensity based on the specific requirements of stored goods.
- **Real-time Temperature Monitoring:** Cold chain storage facilities are now equipped with advanced sensors that continuously track temperature levels.

These sensors transmit real-time data to centralized monitoring systems, enabling swift corrective actions if any deviations occur.

- **Automated Cooling Systems:** Automation has revolutionized cold storage management by enabling cooling systems to self-adjust based on factors such as product type, external temperature fluctuations, and storage conditions. This ensures stable temperatures while minimizing human intervention and potential errors.
- **Phase Change Materials (PCMs):** PCMs are substances that store and release thermal energy during phase transitions, such as melting and freezing. These materials are widely used in cold chain packaging, particularly for pharmaceutical products, to maintain stable temperatures during transportation.
- **Cryogenic Cooling:** For goods requiring ultra-low temperatures—such as specialized pharmaceuticals—cryogenic cooling is employed. This method utilizes liquid nitrogen, carbon dioxide, and other cryogenic substances to achieve temperatures as low as -150°C.
- **Data Loggers and Remote Monitoring:** Data loggers, which are small devices that continuously record temperature fluctuations, play a crucial role in cold chain transport. Integrated with remote monitoring technologies, these devices allow stakeholders to access real-time temperature records from anywhere in the world, ensuring product safety.
- **RFID Technology:** Radio Frequency Identification (RFID) tags, when combined with temperature sensors, offer dual benefits. They not only provide location tracking but also monitor temperature, ensuring that goods remain within the required range throughout their journey.

Figure: 8.1 Storage and transportation

- **Cloud-based Cold Chain Management:** Cloud computing has significantly enhanced cold chain logistics by offering remote access to temperature data, facility monitoring, and even remote control over certain storage parameters, streamlining the entire process.

- **Advanced Insulation Materials:** Innovations in insulation technology have led to the development of superior materials such as vacuum-insulated panels and aerogels, which enhance temperature stability in cold chain storage and transport.
- **Battery-backed Power Systems:** Power disruptions can pose a severe threat to cold chain operations. Many modern facilities incorporate backup battery systems to ensure uninterrupted cooling, even during outages.

These advanced technologies collectively strengthen cold chain logistics by improving efficiency, reducing waste, and ensuring that perishable goods remain in optimal condition throughout storage and transportation.

GEL PACKS IN COLD CHAIN

Gel packs and ice packs play a crucial role in cold chain packaging by providing additional cooling to temperature-sensitive products. These packs are pre-frozen and strategically placed within the packaging to absorb heat and maintain a stable, cool environment.

In cold chain logistics, gel packs are widely used to keep perishable goods, such as fruits and vegetables, at optimal temperatures during transportation and storage. The cold chain refers to a systematic approach that ensures the proper handling, storage, and distribution of products requiring specific temperature controls.

There are various types of gel packs available, each catering to different applications and needs:

- **Reusable Gel Packs:** Designed for multiple uses, these gel packs can be refrozen and reused across multiple shipments. They are ideal for businesses that use the same packaging multiple times, reducing waste and operational costs.
- **Disposable Gel Packs:** Intended for single-use shipments, these packs provide a convenient cooling solution for deliveries where the cold chain is maintained for only one journey. They eliminate the need for retrieval and reuse, making them suitable for long-distance shipping.
- **Flexible Gel Packs:** These packs are designed to conform to the shape of the packaged products, ensuring even cooling and maximum surface contact. Their adaptability makes them useful for shipments with irregularly shaped items.
- **Hard Gel Packs:** More rigid in structure, these gel packs retain their shape even when frozen, providing a consistent cooling surface within the packaging. They are particularly effective in maintaining uniform temperatures for extended periods.

By incorporating the right type of gel packs, businesses can enhance their cold chain operations, ensuring the safe and efficient transport of perishable goods while minimizing temperature fluctuations.

How are Ice Gel Packs Operational?

Ice gel packs function based on thermodynamic principles, utilizing the latent heat of fusion to maintain temperature stability. When frozen, the gel inside transitions from a liquid to a solid state, absorbing a substantial amount of heat from its surroundings and effectively lowering the temperature.

These gel packs play a crucial role in absorbing and releasing thermal energy, acting as a buffer against temperature fluctuations. As external temperatures rise, the packs absorb excess heat, preventing the contents from warming too quickly. Conversely, when temperatures drop, they release stored cold energy, counteracting freezing conditions and maintaining an optimal environment for temperature-sensitive products.

Cold Chain Packaging with Ice Gel Packs

Proper handling and transportation practices are crucial to maximizing the effectiveness of ice gel packs in maintaining a stable temperature. Strategically placing gel packs within the packaging ensures uniform cooling, preventing extreme fluctuations throughout transit.

Achieving Even Cooling

To maintain a consistent temperature, distribute gel packs evenly inside the packaging. This prevents localized hot spots and ensures that all contents receive adequate cooling. Positioning gel packs near the outer edges helps sustain a stable internal temperature.

Preventing Direct Contact

While gel packs effectively regulate temperature, direct contact with products can lead to unintended freezing, which may damage delicate items. To mitigate this risk, use protective barriers such as cardboard dividers, foam spacers, or insulation layers to create a buffer between the gel packs and the products.

Guidelines for Safe Handling

Proper handling is essential to prevent damage to the gel packs and maintain their cooling efficiency. Avoid dropping, crushing, or puncturing the packs, as leakage can compromise the cooling system. Clearly label packages containing gel packs to ensure careful handling during transportation.

Maintaining Gel Pack Integrity

Regularly inspect gel packs throughout the shipping process to ensure they remain intact and functional. Any leakage or damage can lead to temperature inconsistencies, potentially affecting the quality of the stored items. Visual checks help identify compromised gel packs before they impact product safety.

Layering and Insulation

Layering and insulation create a barrier that slows the exchange of heat between the external environment and the contents of the package. This barrier is essential to maintaining the cold chain and ensuring that the gel packs have sufficient time to effectively regulate temperature.

Maximize cold chain packaging efficiency by strategically using insulation and layering techniques. This approach not only helps protect temperature-sensitive contents, but also helps minimize waste and preserve product quality during transport:

- **Proper Positioning in Packaging:** Ensure that gel packs remain correctly positioned in the packaging during transport. Vibration, shock, and changes in orientation can move gel packs out of their intended location. This movement can result in uneven cooling or inadequate temperature control within the package.
- **Improve Temperature Retention Through Insulation:** Insulating materials, such as bubble wrap, foam, or thermal blankets, protect against temperature fluctuations. These materials trap air and act as barriers to heat transfer. By providing the interior walls of the package with insulation, you minimize the impact of external temperature changes on the contents. This is especially important when products are exposed to varying temperatures during shipping or storage.
- **Creating a Cold Barrier Through Layers:** Strategic placement of gel packs and products can help create a cold barrier that prevents temperature changes. Place a layer of products on the bottom, followed by a layer of ice packs. Repeat this process alternating products and gel packs until the pack is full. This method ensures that products are surrounded by cooling sources on all sides, reducing the risk of hot spots or insufficiently cooled areas.

Monitoring Temperature

Proper cold chain operation relies primarily on careful monitoring of temperature levels within the packaging. Temperature-sensitive indicators are essential to keep a close eye on this critical aspect:

- **Maintain Consistency in Temperature Regulation:** Gel packs work in synergy with the packaging design to regulate temperature. Any misuse or incorrect positioning can disrupt this delicate balance. To maintain consistency in temperature regulation, emphasize the importance of good handling practices for all actors involved in the transportation process.
- **Temperature-sensitive Indicators:** Integrate temperature-sensitive indicators or data loggers into the package. These small devices provide real-time data on internal temperature. Choose indicators that are consistent with the temperature range required for specific products. These indicators provide valuable information about temperature deviations and help you take quick corrective action.
- **Regular Checks and Record Keeping:** During transport, maintain a schedule to check and record temperature readings. Regularly evaluate the performance of the gel pack and the overall temperature stability. This practice helps to identify fluctuations or problems that may occur during transportation. Keep

detailed records of temperatures, which can be extremely useful for analyzing and improving cold chain processes.

Maximizing Gel Pack Performance

Maximizing the performance of ice gel packs requires meticulous planning and execution to ensure steady and effective cooling throughout transportation.

Avoid Over-Packing for Better Airflow

Refrain from overcrowding the packaging with too many gel packs, as excessive packing can obstruct airflow, reducing their overall cooling efficiency. Proper air circulation is essential for gel packs to function optimally by effectively absorbing and releasing heat. Without sufficient space, heat exchange is restricted, leading to inconsistent temperature control. Ensuring adequate spacing allows for better air movement, enhancing cooling performance.

Strategic Use of Different Gel Pack Sizes for Uniform Cooling

Utilizing gel packs of various sizes strategically within the package helps fill gaps and creates a more balanced cooling effect. Smaller gel packs can be placed in compact spaces or corners, ensuring that all areas receive adequate cooling. This method prevents temperature variations and enhances the even distribution of cold energy, maintaining a stable internal environment for temperature-sensitive products.

Dry Ice

Dry ice is extremely cold, with temperatures ranging from -78.5 to -109.3°C. Initially, it was considered ideal for transporting perishable products such as vaccines, sensitive medicines, ice creams, and other frozen food items. However, due to its emission of high levels of carbon dioxide, it was eventually replaced by phase change materials (PCM) as a more effective solution.

Maintaining a consistent temperature range is essential when transporting pharmaceutical products like vaccines, biological samples, and other medical supplies. From production through to final use, ensuring the preservation of these products throughout the entire supply chain is critical.

Dry ice has long been a reliable and efficient cooling solution for medical and pharmaceutical shipments within the cold chain. With its impressive cooling capacity, it sublimates directly into gas without leaving behind any residue, guaranteeing consistent temperatures for the products and preventing refrigerant-related damage. Similarly, in the food industry, dry ice is commonly used to package and transport frozen or refrigerated products. It provides an efficient and dependable solution for maintaining the quality and freshness of these goods throughout their journey in the supply chain.

Liquid Nitrogen

This technology is primarily used to keep packages frozen for extended periods and is especially crucial for transporting biological items such as organs, tissues, or bodies. The extremely low freezing point of -196°C ensures that these sensitive items are preserved throughout their journey, maintaining their integrity during transit.

Eutectic Plates

Commonly referred to as cold plates, these are similar to gel packs but use liquid instead of gel, making them reusable multiple times. Cold plates serve a variety of purposes, including maintaining a stable temperature in delivery vehicles and ensuring consistent temperatures in rolling refrigerated units.

Reefers

This can refer to a semi-trailer, van, or ISO container. The term "reefer container" is commonly used for a refrigerated container. These units are specifically designed with special insulation and a refrigeration plant to maintain temperature-controlled air circulation.

National Policy for Cold Chain Standard

The cultivation of fruits and vegetables, along with the establishment of a robust national industry, represents intricate policy challenges that require a comprehensive and integrated approach. National policy should encompass food available in domestic markets—whether locally grown or imported—as well as food intended for export. This ensures that all produce reaching consumers remains safe and of high quality. Fruits and vegetables sold both domestically and internationally must adhere to national regulations and controls, as well as meet the standards set by importing countries.

Additionally, national policy should encourage the consumption of domestically grown fresh produce through well-developed markets and supply chains, stringent quality and safety measures, and the adoption of modern technological advancements as necessary.

A well-structured national policy may focus on key aspects such as ensuring stable market supply, implementing effective post-harvest technologies, optimizing transportation methods, organizing a skilled agricultural workforce, and maintaining strict food safety and quality controls at each stage of the supply chain. A streamlined and organized fruit and vegetable industry facilitates the enforcement of food safety regulations, making it easier to uphold quality standards.

The process of establishing food quality and safety programs in a producing country follows a structured approach. Initially, collaboration with domestic stakeholders is essential to develop standards, which eventually pave the way for either mandatory legislation or voluntary compliance guidelines. The successful implementation of food safety programs is most feasible when the fruit and vegetable sector is strong, resilient, and actively supported by government efforts such as inspections and analytical evaluations. To further enhance international trade, governments can contribute by reinforcing food safety programs within the industry, providing training and technical insights on regulatory compliance, and establishing a robust oversight system capable of conducting thorough inspections, analyses, and certifications.

Recent achievements in integrating food quality and safety measures into the production process have largely stemmed from enhanced coordination among stakeholders through contract farming and the creation of commodity chains. For instance, direct agreements between farmers and hotels or supermarkets strengthen supply chains and reinforce accountability among all parties involved. This interconnected approach is a major factor contributing to its effectiveness.

To ensure the seamless integration of food quality and safety measures across the supply chain, the training and assistance provided to stakeholders should align with their specific needs and production systems. Farmers, as the primary producers, have a crucial responsibility in prevention, which is addressed through the adoption of Good Agricultural Practices (GAPs) and Good Handling Practices (GHPs). In many developing countries, small-scale farmers require government support to uphold food safety and quality standards.

For such measures to be truly effective, the responsibilities of farmers in producing safe, high-quality, pest-free food must be considered alongside other critical factors that affect marketability. These include crop selection, transportation logistics, and accessibility to markets. Food safety education initiatives should be harmonized with broader efforts aimed at enhancing agricultural productivity and market reach.

Food Safety Standards: HACCP

HACCP, which stands for Hazard Analysis and Critical Control Points, is a systematic approach designed to ensure food safety. This method governs how food is processed and handled across Europe and was officially incorporated into the German Food Hygiene Ordinance in 1998. A fundamental aspect of this system is the identification of potential hazards and the critical control points where intervention is necessary to maintain food safety standards.

One of the key principles of HACCP is maintaining proper temperature control during food transportation and storage. To comply with HACCP guidelines, temperatures specified on food packaging must be routinely monitored. Furthermore, clear minimum and maximum temperature limits must be established to prevent food spoilage. A crucial requirement of HACCP is the continuous documentation of all temperature readings to provide proof that food is consistently kept at the correct cold storage conditions.

The appropriate temperature settings for transporting food are determined either by regulatory guidelines or by the food manufacturer's recommendations. If the food is pre-packaged and ready for consumption, the manufacturer specifies the ideal storage temperature. However, for raw ingredients, legal regulations dictate strict temperature limits to ensure food safety. For example, raw chicken must always be stored at or below +4°C, while minced meat should never exceed +2°C to prevent bacterial growth and contamination.

FOOD SAFETY AND STANDARDS ACT, 2006

The Government of India passed the Food Safety and Standards Act, 2006 (FSSA, 2006), a comprehensive law that governs and monitors food safety and quality standards throughout the nation. The purpose of the Act was to create a single statutory body in charge of guaranteeing food safety in India, streamline food regulations, and combine disparate food laws. The Food Safety and Standards Authority of India (FSSAI) was established in accordance with the provisions of this Act.

Objectives of the Act

The FSSA, 2006's main goal is to guarantee that food products on the market adhere to scientific safety standards and do not endanger public health. The Act's main objectives are consumer protection, preventive food safety measures, and bringing Indian food laws into compliance with international standards. It guarantees that consumers have access to high-quality food products by promoting hygienic processing, safe food handling procedures, and clear labeling.

Main Features of the Act

Establishment of FSSAI

This Act established the Food Safety and Standards Authority of India (FSSAI) as the highest regulatory body responsible for establishing food safety guidelines, carrying out surveillance, and keeping an eye on food enterprises.

Uniform Food Laws

The Act unified all food regulations under a single framework and superseded several food-related laws, such as the Prevention of Food Adulteration Act, 1954.

Scientific-Based Standards

The Act requires that international best practices, risk assessment, and scientific research serve as the foundation for food safety standards.

Regulation of Food Business Operators (FBOs)

All food producers, importers, distributors, and vendors are required to secure FSSAI licenses and adhere to stringent safety regulations for the processing, handling, and storage of food.

Consumer Rights and Penalties

By requiring accurate labeling, ingredient transparency, and nutritional information, the Act protects consumer rights.

Penalties, fines, or incarceration may result from noncompliance, contingent on the seriousness of the offense.

India's food safety standards have been greatly raised by the historic Food Safety and Standards Act, 2006. The Act is essential to safeguarding public health and encouraging safe food practices across the country because it unifies food laws and ensures stringent regulatory oversight.

Food Safety and Standards Act, 2006 - Amendments

To improve compliance procedures, fortify food safety regulations, and bring Indian food laws into line with international standards, the Food Safety and

Standards Act, 2006 (FSSA, 2006) has undergone a number of amendments. The main goals of these amendments are to strengthen penalties, streamline licensing, and add new safety precautions. The Food Safety and Standards (Amendment) Bill, 2020, which sought to strengthen food testing infrastructure, enhance supply chain traceability, and impose harsher penalties for food adulteration, was one of the main amendments. For those found guilty of serious food adulteration, it imposed harsher penalties, such as life in prison and hefty fines.

Simplifying the licensing and registration procedures for Food Business Operators (FBOs) was another important amendment pertaining to ease of doing business. Additionally, it concentrated on using online portals to digitize compliance mechanisms, increasing the efficiency and transparency of regulatory procedures.

Furthermore, by strengthening its regulatory authority and broadening its scope in food research and surveillance, the amendments have reinforced the Food Safety and Standards Authority of India (FSSAI). The government's dedication to upholding strict food safety regulations while promoting business expansion and consumer protection in India is reflected in these amendments.

9

Supply Chain Management Systems Planning

Supply chain management (SCM) is essential for the food industry because it involves the coordination and management of various activities from food procurement to consumption. The food supply chain is complex and involves many stages, from the production of raw materials to packaging, distribution and sales. Therefore, effective food supply chain management is essential to ensure that food products are delivered to consumers efficiently, cost-effectively and with maximum food safety.

One of the biggest challenges in food chain management is food safety. Foodborne illness is a major concern and food manufacturers must implement strict quality control measures at every stage of the food supply chain. This includes testing and monitoring for pathogens, maintaining proper storage and transportation conditions, and ensuring that all equipment is cleaned and disinfected to prevent contamination.

Another challenge is food waste. Food products have a limited shelf life and Managing inventory levels is essential to minimize waste and spoilage. Effective food supply chain management requires careful planning and coordination between suppliers, manufacturers, distributors, and retailers to ensure that products are delivered to consumers before their expiration date.

Efficient management of logistics and transportation is another essential aspect of supply chain management. Food products often travel long distances before reaching consumers, and companies need to ensure that products are delivered to customers quickly and efficiently. This requires careful planning and coordination of transportation routes, scheduling, and delivery times to minimize delays and disruptions.

The food industry also relies on effective communication and collaboration between all parties in the supply chain. Setting clear expectations, sharing information and data, and working together to identify and resolve emerging issues or challenges are essential to ensuring that all parties work together to deliver high-quality food products to consumers.

Technological advances have also had a significant impact on supply chain management. Using sensors and other IoT devices can help companies monitor temperature, humidity, and other environmental conditions during transportation and storage, helping to ensure that food products are delivered in optimal conditions. Data analytics and machine learning can also be used to improve demand forecasting and planning, helping companies better manage inventory levels and minimize waste.

Effective food supply chain management is essential to ensure food safety, minimize food waste, and deliver high-quality food products to consumers. Food manufacturers must implement strict quality control measures, manage inventory levels, and maintain efficient logistics and transportation management to ensure that products are delivered on time and efficiently. Effective communication and collaboration between all parties in the food supply chain are also essential to achieving these goals. By using the latest technologies and working collaboratively, food companies can ensure they deliver high-quality products to consumers in the most efficient and cost-effective way possible.

FRUIT SUPPLY CHAIN MANAGEMENT IN INDIA

Recent developments in global food markets are driven by consumer demand and preferences, concerns about food safety, and the increased bargaining power of modern retail systems. Rising incomes and changing lifestyles have led to a demand for fresh, "healthy," and convenient products with better quality and a more diverse offering throughout the year. In addition, consumer concerns about food safety and the social and environmental conditions under which food is produced have prompted us to think. that technological interventions are needed to enhance supply chains and product development. Therefore, it is important to preserve the value of important perishable goods at every stage of the supply chain, as well as during processing.

The traditional fruit and vegetable supply model in India (Figure 9.1) From producer to consumer, there are a number of intermediaries. Quantitative losses due to physiological losses and quality losses due to poor handling occur at every stage. In some parts of the country, racks of bananas are stacked on top of each other and transported in open carts and mini-trucks to the auction/wholesale markets. Since the country's roads and access roads to the auction/wholesale markets are undulating, bumpy and with very few roads, the deterioration of the fruit has intensified. Losses accumulate at each stage of the supply chain (Figure 9.1) and a long supply chain like that used in the conventional system results in a significant loss of pH and fruit quality.

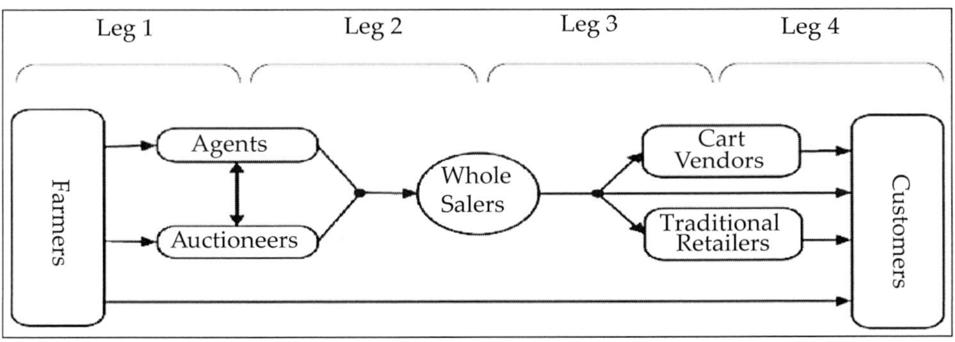

Figure: 9.1 Traditional supply chain system followed in India

In various regions, bamboo baskets lined with paddy straw cushioning are commonly used for packaging fruits like papayas and mangoes. However, this traditional packaging method leads to significant spoilage during storage and transportation, drastically impacting the quality of the fruit. The resulting losses diminish the market value of the produce for farmers, often forcing them to engage in distress sales. For most farmers, the traditional supply chain used for fresh fruits rarely offers adequate financial returns, leaving them with limited profits.

Figure: 9.2 Packaging of mangoes in bamboo baskets with paddy straw as cushioning material

The supply chain followed by organized companies like Reliance Fresh and Aditya Birla Retail Ltd (More) in India is notably shorter than the traditional model. These companies directly source produce from farmers, farmer producer organizations (FPOs), farmer co-operatives, or farmer clusters. In many cases, the sorting of fruits takes place at the farm gate, with only the desired produce being selected for retail distribution.

The supply chain model involves transporting fresh fruits and vegetables to collection centres (CC) or buying centres, where they are weighed, sorted (if needed), and then moved to distribution centres (DC). At the DCs, primary processing operations like grading, packaging into small retail packs are conducted. From there, the produce is sent either to retail stores of the companies or directly to consumers via e-commerce platforms. The DCs are equipped with facilities such as graders, conveyors, washers, and cold stores to maintain the quality of the products.

Reliance Fresh utilizes a fleet of refrigerated vehicles to transport fruits to its retail stores, significantly reducing post-harvest (PH) losses. According to Sihariya et al. (2013), Reliance Fresh has successfully reduced these losses from 25-30% at the farm gate to just 7-8% by implementing precooling of the harvest, improving post-harvest handling (such as reducing human touch/contact), and using specialized packaging for highly perishable products. Additionally, Reliance utilizes cold chain logistics for the inter-state transportation of fruits and vegetables and conducts regular staff training within its supply chain to further reduce PH losses.

Way Cool Foods and Products Pvt Ltd, located in Chennai, sources most of its fruits and vegetables directly from farmers and FPOs in Chennai and surrounding areas in Tamil Nadu. The company then transports fresh produce to its collection centre, which is equipped with facilities for sorting, grading, weighing, washing, and temperature-controlled compartments for storing highly perishable items. The company repackages the produce into bulk packages for wholesalers and smaller retail packages for its franchisee retail outlets. Additionally, Way Cool uses Phase Change Material (PCM) technology in its transport vehicles to maintain temperature control while transporting perishable products. To further reduce PH losses at the retail level, the company has installed ICAR-IIHR developed Arka High Humidity storage boxes in its franchisee outlets, significantly improving storage conditions for green leafy vegetables and other perishables.

Big Basket, an e-commerce giant supplying 100,000 tons of fresh fruits and vegetables annually, directly procures around 80% of its produce from farmers and FPOs, with the remaining produce sourced from companies like Grofers and Way Cool. The company operates collection centers (CCs) located near production hubs, equipped with basic facilities for weighing, sorting, and grading produce. Afterward, fresh produce is transported to distribution centers (DCs), where it is repackaged into smaller retail packs (such as punnets for pomegranates, CFB boxes for mangoes and papayas), then delivered directly to consumers' doorsteps. Some DCs also include facilities for fresh-cut produce processing and packaging. For highly perishable items, Big Basket uses refrigerated vans or containers with gel packs for transporting small retail packs. While these companies have significantly reduced PH losses and added value to the fresh horticultural produce, they still collectively source less than 5% of the total fruits produced in the country.

MANAGING THE SUPPLY CHAIN FOR FRESH PRODUCE

In the fresh produce industry, managing the post-harvest supply chain is particularly challenging. Mistakes are not forgiven as easily as in most other industries. A shipment of produce left in a warehouse for a week is not a concern, it is a disaster. The entire load can be lost.

Managing the fruit supply chain represents an even greater challenge. In the enclosed, highly compact environments of fruit storage, a major problem is the production of ethylene, the agent that ripens fruit. Without proper management, ethylene can build up quickly and cause accelerated decomposition and waste.

The key to effective supply chain management is to look for opportunities to make these types of improvements wherever possible. Add a chain of improvements and the positive effect can be transformative.

A Effective supply chain management is a critical part of any business, having a direct and measurable impact on sales, revenue, profits, customer satisfaction and much more. In the world of fresh produce, especially, this is a managerial responsibility that you ignore at your peril.

The Supply Chain Umbrella

A wide range of activities beyond purchasing plays a crucial role in supply chain management. These activities, though diverse, are all interconnected and define the efficiency and effectiveness of how goods and information move across a supply chain. The key activities include:

- **Purchasing**: Purchasing is a central activity in most organizations, forming the core focus of supply chain operations.
- **Quality Control**: Every organization understands the importance of maintaining supplier quality, and many focus on preventing quality issues rather than just detecting them. Leading organizations collaborate with suppliers to develop strong quality control processes and procedures.
- **Demand and Supply Planning**: Demand planning involves forecasting the anticipated demand for products, adjusting inventories, tracking unfilled orders, and managing spare parts or after-market requirements. Supply planning translates this demand data into a robust supply, production, and logistics network that can meet demand.
- **Material or Inventory Control**: The material group typically oversees determining the required inventory levels of finished goods to meet customer demand, focusing on the physical distribution (outbound or downstream) aspect of the supply chain. The inventory control team plays a vital role in managing these inventory levels.
- **Order Processing**: This activity ensures customers receive materials at the right time and place. It serves as a critical link between the producer and external customers, helping to fulfill customer orders effectively.
- **Production Planning, Scheduling, and Control**: Involves developing time-phased production schedules, creating short-term plans, and overseeing the management of work-in-process production to ensure the manufacturing process runs smoothly.
- **Warehousing/Distribution**: Particularly significant for companies that produce goods based on forecasts, warehousing and distribution help manage inventory and ensure products are available when required for future sales.
- **Customer Service**: Customer service encompasses a broad range of activities aimed at ensuring customer satisfaction with both products and services.

Supply Chain Planning

When people think of supply chain planning, the focus often lands on inventory and logistics. However, it's far broader than that. For physical products and manufacturing, it begins long before the product reaches the shelf or consumer door. It starts with managing suppliers who source raw materials and extends all the way through to delivery, and even beyond, with returns, recycling, and reverse logistics. Consumer behavior also plays a crucial role at every stage: their shopping patterns, feedback, and ever-evolving preferences.

Supply chain planning involves forecasting demand and planning for the necessary materials and components to meet that demand. It includes aspects like production, marketing, distribution, and sales.

The first part of this definition focuses on anticipating demand for products. If you have an accurate system in place for demand forecasting, much of the challenge in supply chain planning is already addressed.

The goal of planning is to balance supply and demand, ensuring that sales opportunities are maximized in a timely manner.

A rational approach in supply chain processes is to begin by looking at the desired outcome and then work backwards. This principle applies to supply chain planning, which typically starts with demand planning.

Teams from various departments—such as sales, marketing, and production—collaborate to estimate demand based on historical data and other factors. Once demand is projected, it's passed on to production, logistics, and other relevant teams. Anticipating demand helps maintain optimal inventory levels, avoiding the costs associated with overstocking.

Elements of Supply Chain Planning Process

For supply chain planning to be effectively executed, seamless coordination across various teams and processes is essential. Each part of the chain needs to work together to ensure smooth operations and timely deliveries.

Below are the key elements of supply chain planning:

Demand Planning

As mentioned earlier, demand planning forms the backbone of the entire supply chain process. It involves evaluating diverse data to forecast demand accurately, which in turn helps maintain optimal inventory levels and avoid excess stock, thereby reducing associated costs. The process of demand planning incorporates historical data, projected sales, market trends, and other relevant factors.

Today, demand planning is increasingly driven by advanced technology. With the power of artificial intelligence (AI) and machine learning (ML), modern supply chain management software is able to predict demand with high precision. By analyzing real-time data, market conditions, and point-of-sale information, these systems can 'sense' demand more accurately.

Additionally, the use of predictive analytics has become more prevalent in demand forecasting. This allows businesses to gain deeper insights into consumer behavior, buying patterns, and other factors that influence demand, enabling more effective planning and decision-making.

Supply Planning

The subsequent step is to create a supply plan that aligns with the demand plan and fulfills the overall business requirements.

The supply plan focuses on sourcing the necessary raw materials, components, and other goods required for production. The aim is to ensure that the supply of these materials is efficiently coordinated to meet product demand in the most optimal way.

Production Planning

This involves the allocation of resources, including employees, materials, and production capacity. The primary goals of a production plan are to minimize waste and produce only what is necessary, ensuring that there is optimal inventory availability. Achieving the latter is made possible through a well-organized supply chain plan that enhances cross-functional visibility and supports efficient inventory management.

Sales and Operations Planning

Typically conducted on a monthly basis, Sales and Operations Planning (S&OP) brings together various business teams with different goals, aligning them toward a common purpose. It allows sales and marketing leaders to evaluate and integrate their plans with operations.

This collaborative meeting of diverse teams ensures the coordination of supply assets and resources to meet both short-term and long-term demand needs.

In some organizations, S&OP is part of a broader process called integrated business planning. This process includes the incorporation of other departments, such as finance and HR, into a unified, company-wide plan.

Advantages of Supply Chain Planning

Supply chain planning can help streamline operations, improve efficiency, avoid delays, and reduce operating costs. Done well, planning can reduce costs associated with many processes, including purchasing, manufacturing, inventory management, and logistics.

In addition, planning can improve customer satisfaction by ensuring that products are available in the right place at the right time. Increased customer satisfaction, in turn, helps the company build customer loyalty.

In today's business environment, it is imperative that companies of all sizes engage in supply chain planning.

You need to establish a comprehensive plan that synchronizes demand, supply, production, transportation, and distribution. In addition, it should also cover risks and include contingency plans for common points of failure.

Without a comprehensive supply chain plan, a company lacks the foresight and preparation to effectively address potential scenarios.

To Design a Supply Chain Planning System

So far, we have discussed the significance of supply chain planning and the components necessary for it, but without a clear understanding of how to assemble these elements, they remain ineffective. Here are the key steps to consider:

- **Collect Data:** The more high-quality data a manufacturer can collect and analyze, the more precise its planning models and forecasts will be. This data includes past sales figures, current budgets, inventory levels, manufacturing capacity, and resources, along with data from Sales and Operations Planning (S&OP). Additionally, data on market size, customer sentiment, buying habits, economic conditions, and other relevant factors are important. To gather external data, manufacturers may use third-party aggregators, research institutions, journals, and reputable industry publications. Scenario planning software can simplify this by automatically collecting external data. Internal data can be gathered manually or through supply chain planning software, which automates the process. The final step is combining both external and internal data so that planners can assess how various scenarios might impact the supply chain.
- **Adopt Just-In-Time (JIT) Principles:** Originally developed in Japan post-WWII, with Toyota being a major adopter, JIT is a lean inventory strategy that has now been globally implemented. The idea is that by ordering raw materials and components just in time for processing, manufacturers can reduce inventory levels and lower related costs. However, effective supply chain planning also involves considering contingency plans in case of supply disruptions or unexpected demand increases. For these scenarios, manufacturers typically keep a certain amount of safety stock on hand.
- **Prioritize Visibility:** Visibility within the supply chain refers to tracking each material, product, and service throughout its journey. Manufacturers must ensure all parties involved are aligned, ideally using a digital system that everyone can access and use to provide updates. Cloud-based applications help with visibility as authorized individuals can access them from any location. Manufacturers must also set clear quality expectations and ensure these standards are visible to all involved parties, managing them from the start to prevent discrepancies in quality from one step to the next. Enhanced visibility allows manufacturers to quickly identify and resolve issues such as inventory shortages or logistical problems.
- **Standardize Processes:** Consistency in processes is essential for all stages of the supply chain, from product planning to final delivery. Standardization goes beyond automating the assembly line, although this is often an initial step. In today's supply chains, standardization frequently involves using the same software across all manufacturing locations and ensuring that suppliers have access to the same information. This ensures that production lines follow a uniform process every time, including ordering materials, issuing purchase orders, logistics, quality control, assembly, and sales. Standardized, automated processes allow for economies of scale, reduce human error, and facilitate compliance with quality control and regulatory standards. Without standardization, information silos are created, which hinders the ability to view the entire supply chain and accurately develop a plan to meet demand.

Supply Chain Management Challenges

There are three common problems in supply chain planning. These problems are generally observed in the supply chain management system as a whole; They also apply to subcategories, such as planning:

- The first problem is the minimal organizational commitment to planning development. It is often difficult to bring together the different stakeholders of a business process on the same platform.
- Another problem is the lack of collaborative efforts between different functions, which leads to obstacles in the creation of plans.
- When evaluating supply chain planning, companies tend to look at historical data collected in their systems. However, there is much more dynamic data available that can help companies better prepare and improve their operations.

TYPICAL SUPPLY CHAIN MANAGEMENT ISSUES

Supply chain management plays a crucial role in every industry, particularly in the food sector. However, businesses in the food industry face several hurdles when managing their supply chains. This section discusses some of the common challenges encountered in food supply chain management. One of the primary challenges in managing the food supply chain is ensuring food safety. The food industry is subject to strict regulations to guarantee that food products are safe for consumption. Several factors can affect food safety, such as contamination during production or transportation, improper storage conditions, and insufficient packaging. Supply chain managers must collaborate closely with suppliers and distributors to ensure food products are handled and stored correctly while being transported under safe conditions. Additionally, they must implement rigorous quality control processes to ensure products remain free from pathogens and other contaminants.

Another issue in food supply chain management is the complexity of the global food network. Many food products are grown in one part of the world and then shipped across the globe for processing and packaging. This complexity can make it difficult for supply chain managers to monitor the entire process and ensure proper handling of food products. Furthermore, international regulations and cultural differences can add challenges for businesses trying to navigate the global food supply chain.

Supply chain managers must also contend with challenges like fluctuations in demand and supply chain disruptions. Demand fluctuations can occur due to seasonality, sudden changes in consumer preferences, or unexpected events, such as pandemics. Disruptions in the supply chain, such as natural disasters, transportation strikes, or supplier bankruptcies, can also impact the timely delivery of food products. When these disruptions arise, supply chain managers must act swiftly to ensure food products are still delivered promptly and safely to consumers.

Managing inventory is another key challenge in the food supply chain. It's essential to maintain optimal inventory levels to ensure on-time delivery while minimizing waste. Striking the right balance is difficult: overstocking can result in spoilage, while understocking can lead to stockouts and missed sales. Supply chain managers use data analytics and forecasting techniques to optimize inventory levels, ensuring food products are always available to meet customer demand.

Sustainability is also an increasing concern in the food industry. Consumers are becoming more conscious of the environmental impact of food production and transportation, demanding sustainable options. Supply chain managers must collaborate with suppliers and distributors to ensure food products are produced and transported in an eco-friendly manner. They must also find ways to reduce waste and minimize the carbon footprint of the food supply chain.

Effective supply chain management is crucial to the food industry's success, yet it presents many challenges that require careful planning and coordination. From ensuring food safety and managing inventory to responding to disruptions and embracing sustainability, supply chain managers must navigate a range of issues to ensure timely, safe, and efficient delivery of food products. By adopting advanced technologies, working closely with suppliers and distributors, and implementing strict quality control measures, businesses can overcome these challenges and build a robust and efficient food supply chain.

INDIAN SUPPLY CHAIN MANAGEMENT SOFTWARE

For companies in a variety of industries, supply chain management (SCM) software is essential to maximizing inventory, logistics, and operational effectiveness. Advanced supply chain management (SCM) solutions that improve visibility, streamline operations, and cut costs are in high demand in India due to the country's fast growth in e-commerce, manufacturing, retail, and agriculture. Indian supply chain management software ensures smooth coordination between manufacturers, distributors, and suppliers by addressing the particular difficulties of the nation's intricate and varied logistics network.

Features of Indian SCM Software

1. **End-to-End Visibility:** Effective supply chain management is ensured by real-time tracking of inventory, shipments, and warehouse operations.
2. **Demand Forecasting:** AI-powered analytics assist companies in anticipating patterns in demand and maximizing stock levels.
3. **Multi-Modal Logistics Support:** To guarantee efficient logistics operations, integration with air, sea, rail, and road transportation systems is necessary.
4. **Compliance and Documentation:** Operations are made simpler by automated adherence to FSSAI standards, GST regulations, and other Indian trade laws.
5. **Cold Chain Management:** This is crucial for sectors like food and pharmaceuticals, as it guarantees the safe storage and transportation of goods that are sensitive to temperature.

Popular Indian SCM Software Solutions

Numerous domestic SCM software companies have become important participants in the market. Among the noteworthy ones are:

- **FarEye:** A platform for supply chain visibility and logistics that improves the effectiveness of last-mile deliveries.
- **LogiNext:** A data-driven logistics automation program that maximizes warehouse and transportation operations.
- **Unicommerce:** A cloud-based system for managing orders and inventory for retail and e-commerce companies.
- **Shiprocket:** focuses on helping small and medium-sized businesses (SMEs) automate their shipping procedures.
- **Delhivery:** A supply chain and logistics platform powered by technology that provides transportation and warehousing options.

SCM software is now necessary for businesses to stay efficient and competitive in India's quickly growing digital economy. Indian supply chain solutions' use of AI, blockchain, and IoT is spurring innovation and improving the speed, intelligence, and dependability of logistics operations. Indian supply chain management software will be essential to ensuring smooth, economical, and transparent supply chain management as companies continue to embrace digital transformation.

SOURCING IN SUPPLY CHAIN

In fruit and vegetable supply chain management, sourcing involves selecting the right goods and services to purchase. It includes vendor selection, contract negotiation, and supplier performance measurement.

Sourcing, within the realm of supply chain management (SCM) for fruits and vegetables, is a vital process that focuses on identifying, selecting, and managing suppliers that provide fresh produce. This step ensures that high-quality fruits and vegetables are consistently available in the right quantities, at the right time, and at competitive prices. It also addresses challenges like seasonality, quality assurance, sustainability, and logistics. Given the perishable nature of fresh produce, sourcing requires meticulous planning to minimize spoilage and ensure timely deliveries, making it a crucial element in maintaining an efficient and reliable supply chain.

Important Elements of Sourcing in SCM for Fruits and Vegetables

Supplier Selection

- **Local vs. Global Sourcing**:
    - **Local Sourcing**: This approach involves purchasing produce from nearby farms or local suppliers. It helps reduce transportation time, lower costs, and ensure fresher products. Local sourcing is particularly beneficial for seasonal produce, as it offers quicker access to fresh items.
    - **Global Sourcing**: Sourcing from international suppliers provides access to a broader range of fruits and vegetables throughout the year, regardless of local growing seasons. However, it introduces challenges such as longer transportation times and higher logistics costs, particularly for perishable goods.

- **Supplier Criteria**:
  - **Quality Standards**: Producers must adhere to strict quality requirements, which include factors such as appearance, size, ripeness, and the absence of pests or diseases.
  - **Reliability**: Suppliers must consistently deliver produce on time and in the agreed-upon quantities. The reliability of deliveries, particularly in terms of timing and quality, is critical, especially for perishable goods.
  - **Compliance**: Suppliers should comply with both local and international food safety regulations, covering areas such as pesticide usage, certifications, and produce traceability.
  - **Sustainability**: There is an increasing focus on sourcing that promotes environmentally responsible practices. This includes adopting sustainable farming methods, minimizing pesticide usage, and reducing packaging waste.
  - **Contracts and Agreements**: Clear contracts that outline delivery schedules, quality standards, pricing, and penalties for non-compliance are essential for ensuring smooth and efficient operations.

Seasonality and Availability

- **Seasonality Considerations:**
  - The availability of various fruits and vegetables is significantly impacted by the growing season in different regions. Supply chain management (SCM) for fresh produce must adapt to fluctuations in production caused by weather patterns, seasonal harvest cycles, and varying crop yields.
  - For instance, strawberries might be available year-round in warmer climates, but their peak season may occur during the summer in cooler regions.
- **Strategic Sourcing**:
  - **Diversified Sourcing**: To mitigate risks from seasonal shortages or weather-related disruptions (such as droughts or floods), companies often adopt a diversified sourcing approach. This strategy could involve sourcing the same produce from different regions or countries to maintain a consistent supply.
  - **Imports and Exports**: Companies may also collaborate with suppliers in countries with opposite growing seasons.

Quality Control

- **Inspection at Source:** Quality control starts at the farm level, where produce is examined for size, color, ripeness, and any visible damage or disease. This initial inspection ensures that only the highest quality produce is sourced, minimizing the chances of spoilage and returns.
- **Third-Party Inspections and Certifications:** In certain cases, third-party agencies may verify that the produce complies with standards. These

certifications confirm that the produce has been grown under ethical practices and meets safe standards.

- **Cold Chain Management:** For fresh produce, it is essential to maintain the cold chain from harvest to delivery to the consumer. Sourcing suppliers who have reliable cold storage facilities and efficient transportation methods is critical for preserving the quality of the product.

Logistics and Transportation

- **Transportation Considerations:**
    - **Refrigerated Transport**: For highly perishable fruits and vegetables, cold chain logistics, including refrigerated transportation, is crucial to maintaining freshness as the produce moves from farms to warehouses or retail locations.
    - **Lead Time and Delivery Scheduling**: Given the short shelf life of fresh produce, efficient logistics are key to ensuring fast delivery while reducing spoilage. SCM for fresh produce requires tight coordination across sourcing, transportation, and distribution to ensure timely deliveries.
- **Transportation Routes**: Efficiently planning transportation routes and utilizing various modes (e.g., air, sea, road) is essential for reducing both transit time and costs. This is especially critical for international shipments, where air freight may be used for high-value, highly perishable goods like berries, while sea freight is preferred for bulkier, less time-sensitive produce.

Price and Cost Management

- **Pricing Strategy:**
    - **Price Fluctuations**: The prices of fruits and vegetables can vary with the seasons, affected by factors like weather, crop yields, and market demand. Sourcing teams must keep an eye on market conditions and negotiate prices with suppliers to secure the best possible rates.
    - **Cost Optimization**: Logistics costs, including transportation, refrigeration, and handling, can be substantial for perishable goods. SCM teams should focus on optimizing these expenses by improving transportation efficiency, reducing waste, or negotiating long-term contracts with suppliers to achieve cost savings.
    - **Price Volatility Management:** The produce market can experience significant price fluctuations. To manage these uncertainties, companies often use hedging strategies, forward contracts, or long-term agreements with suppliers to stabilize prices and secure better rates.

Sustainability in Sourcing

- **Sustainable Farming Practices**: Many companies are placing a growing emphasis on sourcing from suppliers that adhere to sustainable farming practices. This includes actions such as reducing pesticide use, encouraging crop rotation, conserving water, and minimizing the reliance on chemical fertilizers.

- **Eco-Friendly Packaging**: Companies are increasingly focusing on reducing the environmental impact of their packaging. The use of sustainable materials, such as biodegradable containers or reusable crates, helps reduce waste and supports environmentally responsible practices.
- **Fair Trade and Ethical Sourcing**: Some businesses prioritize sourcing from farmers who follow fair trade principles, ensuring that workers receive fair wages, work in safe conditions, and have access to social benefits, contributing to the well-being of farming communities.

Technology and Data-Driven Sourcing

- **Demand Forecasting**: By analyzing historical sales data, weather patterns, and crop yield forecasts, companies can predict the demand for various fruits and vegetables. This allows for more effective planning in sourcing decisions, reducing waste, and preventing overstocking.
- **Supply Chain Transparency**: Technologies like blockchain are becoming more common in sourcing, offering transparency throughout the supply chain. Consumers can track the origin of their produce, assuring them that it was grown using sustainable and ethical methods.
- **IoT and Smart Sensors**: IoT devices are used to monitor the conditions of stored produce, such as temperature and humidity levels, ensuring that the produce is stored under optimal conditions for freshness and quality.

Risk Management and Contingency Planning

- **Weather and Climate Risks**: Adverse weather events like droughts or floods can significantly affect crop yields and lead to supply shortages. Supply chain management (SCM) teams need to mitigate these risks by diversifying suppliers and maintaining buffer stocks when possible.
- **Geopolitical Risks**: For global sourcing, political changes, trade policy shifts, or regional instability can disrupt the availability of fruits and vegetables. SCM must proactively develop contingency plans to address these risks.
- **Supply Chain Disruptions**: Events such as natural disasters, strikes, or pandemics can disrupt transportation systems, causing delays in shipments. Effective risk management requires backup suppliers and alternative transportation routes to maintain supply chain continuity.

Sourcing in supply chain management for fruits and vegetables is a multifaceted process that involves balancing key factors like quality, cost, seasonality, logistics, and sustainability. Effective sourcing guarantees that fresh, high-quality produce reaches consumers promptly while also tackling issues related to perishability and environmental impact. By selecting the right suppliers, integrating technology, and adopting sustainable practices, businesses can streamline their sourcing operations and enhance the overall efficiency of their supply chains.

MANUFACTURING IN SUPPLY CHAIN MANAGEMENT

Manufacturing plays a crucial role in the supply chain management of fruits and vegetables. SCM encompasses the entire process of converting raw materials into finished goods and delivering them to customers. In the context of SCM for fruits and vegetables, manufacturing refers to the series of activities that occur after produce is sourced and before it reaches the consumer. While fruits and vegetables are agricultural products rather than traditionally "manufactured" items, several processes are involved in transforming raw produce into ready-to-consumer products.

These processes may include food processing, packaging, and value-added transformations aimed at extending shelf life, improving convenience, and satisfying consumer demand. Here's an overview of the essential components of manufacturing in the SCM of fruits and vegetables:

Processing of Fruits and Vegetables

- **Description:** Processing in the context of fruits and vegetables refers to the transformation of raw produce into products that are more convenient for consumers, have a longer shelf life, or can be incorporated into food products. This process can involve washing, cutting, juicing, drying, freezing, canning, or even creating specialized products.
- **Types of Processing:**
    - **Canning and Preserving**: A common method to prolong the shelf life of fruits and vegetables, canning helps maintain product quality for extended periods. For instance, tomatoes are often canned to create tomato sauce or paste, while fruits like peaches and pears can be preserved in jars.
    - **Freezing**: Freezing fruits and vegetables is another effective technique to extend their shelf life while preserving their nutritional value and taste. Common frozen products include peas, berries, and spinach.
    - **Drying**: Dehydrating produce like apples, bananas, and raisins, or vegetables such as tomatoes and mushrooms, creates shelf-stable products. Dried fruits and vegetables are often used as snacks or in cooking ingredients.
    - **Juicing and Pureeing**: Certain fruits, such as oranges, apples, and berries, are processed into juices, purees, or concentrates that can be used for beverages or in products like jams and sauces.
    - **Ready-to-Eat Products**: Some produce is pre-prepared for convenience, such as pre-cut salads or fruit bowls. This allows consumers to purchase fresh products with minimal preparation required.
    - **Fermentation**: Certain fruits and vegetables undergo fermentation to create distinctive products like sauerkraut, kimchi, or pickled cucumbers.

Packaging

- **Description**: Packaging plays a vital role in preserving the freshness, safety, and visual appeal of fruits and vegetables. Additionally, it can be designed to enhance consumer convenience.

Packaging Types

- **Fresh Produce Packaging**: For unprocessed fruits and vegetables, packaging types such as plastic trays, bags, or mesh nets are commonly used, allowing for adequate air circulation. For example, berries are often packaged in clamshell containers to minimize damage during transportation.
- **Vacuum Sealing**: Vacuum sealing is employed to eliminate air and prolong the shelf life of fresh-cut produce or processed items, like frozen vegetables.
- **Modified Atmosphere Packaging (MAP)**: MAP involves altering the air composition inside the package to slow down the ripening process. This technique is frequently used for fruits like apples and leafy greens.
- **Sustainable Packaging**: With the growing emphasis on sustainability, there is an increasing use of eco-friendly packaging materials, such as biodegradable films or recycled materials.
- **Convenience Packaging**: Packaging is also designed for ease of use, such as pre-washed, pre-cut, or ready-to-eat vegetable bags, enabling consumers to use the produce with minimal preparation.

Sorting, Grading, and Quality Control

- **Description**: Sorting and grading are essential manufacturing steps that assess the quality, size, and appearance of the produce. These processes ensure that only the highest-quality fruits and vegetables reach the consumer.

Sorting and Grading

- **Manual Sorting**: In this process, workers manually inspect produce, removing damaged or substandard items to ensure that only the best products proceed to the next phase.
- **Automated Sorting**: Many large-scale operations use automated systems, such as optical sorters, which scan and sort fruits and vegetables based on size, color, and appearance. This method is especially common for produce like apples, potatoes, and tomatoes.

- **Quality Control Measures**: Implementing rigorous quality control processes ensures the product meets specific standards for freshness, taste, and appearance. This may involve testing for ripeness, sugar content (for fruits), and checking for the absence of pesticides or contaminants.
- **Ripeness Management**: Certain fruits, such as bananas and avocados, require specialized handling and ripening processes. These fruits are often stored in controlled environments, such as ripening rooms, to manage the timing of their ripening before they are sold.

Cold Storage and Temperature Control

- **Description**: Cold storage and temperature control are essential to prevent spoilage, especially for perishable fruits and vegetables. Even after processing, it's crucial to maintain proper storage conditions to preserve quality.
- **Cold Storage Facilities:** Produce such as leafy greens, berries, and cucumbers requires consistent refrigeration to maintain freshness. Specialized warehouses or distribution centers with controlled environments are utilized to store these items.
- **Cold Chain Logistics:** Maintaining the cold chain (refrigerated transport and storage) from the point of processing to final delivery is vital for extending shelf life and preventing spoilage. This is particularly important for items that will be shipped long distances or internationally.
- **Controlled Atmosphere Storage:** For some fruits, the storage atmosphere is adjusted to slow down the ripening process. This method is often used for fruits like apples, pears, and bananas, which continue to ripen after they are harvested.

Value-Added Products

- **Description**: Many fruits and vegetables are transformed into value-added products, which add convenience or new experiences for consumers. This can include blending fruits into smoothies, making vegetable-based snacks, or creating sauces, jams, and condiments.

Types of Value-Added Products

- **Fruit Juices and Purees**: Common fruits such as oranges, apples, and berries are processed into juices or purees. These can be marketed as beverages or used as ingredients in various other products.
- **Vegetable-Based Snacks**: Vegetables like carrots, potatoes, and kale are processed into chips, crisps, or dehydrated snacks, offering a convenient, health-conscious alternative.
- **Sauces and Jams**: Fruits and vegetables, including tomatoes, peaches, and strawberries, are transformed into sauces, jams, and marmalades. These products serve as condiments or ingredients in cooking.
- **Frozen Meals and Pre-packaged Dishes**: Ready-to-cook or ready-to-eat meals featuring vegetables (e.g., frozen stir-fries, soups, or stews) are becoming increasingly popular for their convenience and time-saving qualities.

Sustainability in Manufacturing

- **Eco-Friendly Practices:**
    - **Sustainable Manufacturing Practices**: The food industry is placing growing emphasis on sustainability. This includes efforts to minimize waste during processing, such as repurposing byproducts for animal feed or composting. Additionally, companies are focusing on reducing water and energy usage and sourcing ingredients in an environmentally responsible manner.

- **Waste Reduction:**
    - **Waste Minimization Efforts**: To reduce waste, companies are repurposing fruit and vegetable peels or pulp in products like juices and soups, or converting waste into compost or bioenergy. These waste reduction strategies play a key role in helping the food industry meet its environmental sustainability goals.
- **Energy and Water Efficiency:**
    - **Carbon Footprint Reduction**: Many processing plants are concentrating on minimizing their carbon footprint by improving energy efficiency, adopting renewable energy sources, and conserving water, particularly in processes that consume significant amounts of water, such as washing and cooling.

Manufacturing in the supply chain of fruits and vegetables is a complex process involving sourcing, processing, packaging, and distribution. It converts raw agricultural products into convenient forms with longer shelf lives, catering to consumer demand for variety and high quality. The efficiency of these processes plays a vital role in preserving product quality, minimizing waste, and ensuring the produce reaches consumers in optimal condition. Sustainable practices are gaining importance in this manufacturing sector, helping create a more environmentally responsible and efficient supply chain.

DELIVERY IN SUPPLY CHAIN MANAGEMENT

Delivering fresh fruits and vegetables through efficient Supply Chain Management (SCM) involves a well-organized process that ensures the products arrive at the end consumer in the best possible condition. Given that fruits and vegetables are highly perishable, timely delivery, freshness preservation, and minimizing waste are essential. Here's a detailed breakdown of how fresh produce is managed within an SCM framework:

Components of SCM in Fruit and Vegetable Delivery

Sourcing and Procurement

- **Description**: The process starts with sourcing fresh fruits and vegetables from farms, local producers, or international suppliers.
- **Key Considerations:**
    - **Seasonality**: Since many fruits and vegetables are seasonal, procurement strategies must align with harvest periods.
    - **Quality Standards**: Suppliers must adhere to quality guidelines regarding ripeness, size, and overall appearance.
    - **Local vs. Global Sourcing**: Depending on the market, produce can be sourced locally, regionally, or from global suppliers.

Packaging

- **Description**: Proper packaging is crucial to safeguard fruits and vegetables during transit and to maintain their freshness.

- **Key Considerations:**
  - **Temperature-Controlled Packaging**: Certain produce (e.g., berries, leafy greens) must be kept cool, often requiring insulated or refrigerated packaging.
  - **Breathable Packaging**: Materials such as perforated plastic or cardboard allow airflow, reducing the risk of spoilage and condensation.
  - **Protective Materials**: Packaging must prevent bruising or physical damage, particularly for delicate items like tomatoes or peaches.

Transportation

- **Description**: Transportation plays a vital role in SCM for fruits and vegetables, as the produce must arrive at its destination swiftly to preserve its freshness.
- **Key Considerations:**
  - **Cold Chain Logistics**: Refrigerated trucks or containers (cold chain) are essential for preventing spoilage of perishable goods.
  - **Delivery Routes and Timing**: Efficient routing and punctual delivery are crucial to minimizing transit time and ensuring product freshness.
  - **Mode of Transport**: Depending on the distance, delivery may involve road, rail, air, or sea transport. While trucks suffice for local deliveries, air freight or sea freight may be necessary for international shipments.

Warehousing

- **Description**: Temporary storage before distribution is often needed if there are delays or when produce is processed in large quantities.
- **Key Considerations:**
  - **Refrigerated Warehouses**: Cold storage is essential for many fruits and vegetables to maintain their freshness and extend shelf life.
  - **Inventory Management**: Effective warehouse management is crucial for tracking stock, ensuring products are fresh, and minimizing the risk of spoilage.
  - **Order Fulfillment**: Once stored, produce is picked and packed according to customer orders, either for direct delivery to consumers or for distribution to retail outlets.

Last-Mile Delivery

- **Description**: This is the final stage of the delivery process where produce is transported from a warehouse or distribution center directly to the consumer or retail store.
- **Key Considerations:**
  - **Timely Delivery**: Freshness is a top priority, so minimizing delivery time is crucial to ensure the produce arrives at its best.

- **Urban vs. Rural Deliveries**: Urban deliveries may require smaller vehicles to navigate through dense areas, while rural deliveries may necessitate larger trucks for longer distances.
- **Direct-to-Consumer Deliveries**: Online platforms, like subscription boxes or farm-to-table services, require precise scheduling for home deliveries.

Technology and Data Analytics

- **Description**: The modern SCM for fruits and vegetables relies on technology to enhance efficiency and optimize delivery processes.
- **Key Considerations**:
    - **Tracking and Monitoring:** Tools like GPS, RFID, and IoT sensors allow for real-time monitoring of produce, ensuring it remains at the correct temperature and is delivered on time.
    - **Demand Forecasting**: Advanced data analytics can predict the demand for various fruits and vegetables, ensuring the right amount is available at the right time, avoiding both overstocking and shortages.
    - **Blockchain for Transparency**: Blockchain technology can track the journey of produce, offering consumers transparency regarding its origin and handling.

Customer Feedback and Returns

- **Description**: Post-delivery feedback and the handling of returns or complaints are vital components of SCM in the produce industry.
- **Key Considerations:**
    - **Customer Satisfaction**: Regular monitoring of customer feedback ensures that produce quality and delivery times align with customer expectations.
    - **Returns or Spoilage**: An efficient returns process is needed for damaged or spoiled items, either by replacing or refunding the product. Identifying the causes of spoilage can help refine supply chain practices.
    - **Sustainability**: Some companies may adopt sustainable practices like composting or offering eco-friendly solutions for returned or spoiled goods.

Delivery Models in SCM for Fruits and Vegetables

Direct-to-Consumer Delivery

- **Description**: Produce is delivered directly from farmers or distributors to consumers, often through online platforms, subscription boxes, or farmers' markets.
- **Benefits**:
    - **Fresher Produce**: Reduces time spent in transit and handling, ensuring the produce is delivered at its freshest.

    - **Reduced Handling and Transportation Time**: Minimizes the steps between harvest and consumer, leading to a quicker delivery.
- **Challenges**:
    - **Higher Delivery Costs**: Direct-to-consumer delivery can be expensive due to smaller order sizes and individual deliveries.
    - **Limited to Local or Regional Deliveries**: Unless specialized transportation methods are used, this model is often constrained to localized areas.

Retailer Delivery

- **Description**: Fruits and vegetables are delivered to retail stores (supermarkets, grocery stores, etc.), which then sell them to end consumers.
- **Benefits**:
    - **Wide Reach to Consumers**: Retailers have established customer bases and broad distribution networks.
    - **Easier to Handle Bulk Quantities**: Retailers can accommodate larger quantities of produce, facilitating more efficient transportation.
- **Challenges:**
    - **Increased Handling and Potential for Spoilage**: The produce may be handled multiple times as it moves through the supply chain, increasing the risk of spoilage.

Wholesale Delivery

- **Description**: Produce is delivered in bulk to wholesalers who then distribute it to retailers or food service providers.
- **Benefits**:
    - **Economies of Scale for Large Deliveries**: Bulk shipments reduce costs per unit, making it cost-effective for wholesalers.
    - **Efficient for International Shipping**: Wholesale deliveries, especially when shipped in bulk, are more suited for international or long-distance transport.
- **Challenges**:
    - **Produce may Experience Longer Transit Times**: Longer shipping durations, especially for international delivery, could affect the freshness of the produce.
    - **Risk of Surplus or Unsold Stock**: Wholesalers may face difficulties with unsold stock, leading to waste or loss of value.

Third-Party Logistics (3PL) Providers

- **Description**: Companies may outsource transportation and delivery to third-party logistics providers that specialize in refrigerated goods and supply chain optimization.

- **Benefits**:
    - **Expertise in Temperature-controlled Logistics**: 3PL providers specialize in managing the cold chain, ensuring that perishable produce remains fresh throughout transport.
    - **Cost-effective for Smaller Businesses**: Smaller companies may not have the infrastructure for their own delivery systems, so outsourcing to a 3PL provider can be more affordable.
- **Challenges**:
    - **Less Control over Delivery Quality**: Outsourcing can lead to reduced oversight, making it harder to ensure that deliveries meet expectations.
    - **Potential Delays**: If the 3PL provider faces operational issues, such as equipment failure or route disruptions, it could cause delays in delivery.

RETURNING IN SCM

Returns (SCM for fruits and vegetables) is an essential aspect of supply chain management in the fruit and vegetable sector. Unlike many other industries, returns in the fruit and vegetable supply chain are not always related to product defects, but are often related to factors such as spoilage, poor quality, or logistical issues. The returns management process can have a significant impact on customer satisfaction, sustainability, and waste reduction.

Aspects of Returning in SCM

Returns in the fruits and vegetables supply chain are an important part of supply chain management. Unlike many other industries where returns are typically due to product defects, in the produce industry, returns are often driven by spoilage, subpar quality, or logistical challenges.

Key Considerations

- **Spoilage and Poor Quality**: Fruits and vegetables are perishable items and, when mishandled or improperly stored, can deteriorate quickly. Returns often occur when customers or retailers receive produce that has spoiled or is not up to the desired quality standards.
- **Logistical Issues**: Transportation delays, improper handling, or damage during transit can also lead to returns. Issues such as incorrect temperature maintenance or long transit times can result in spoiled or poor-quality produce arriving at its destination.
- **Impact on Customer Satisfaction**: The handling of returns directly affects customer satisfaction. A timely and efficient return process, coupled with quality control measures, helps build trust and customer loyalty.
- **Sustainability and Waste Reduction**: Effectively managing returns can also contribute to sustainability efforts. This involves minimizing waste by repurposing or redistributing returned produce where possible, reducing environmental impact.

- **Waste Reduction**: A structured return management process helps reduce waste by enabling the industry to salvage usable produce and ensure that returned items are handled and disposed of in an environmentally responsible way.

Key Aspects of Managing Returns in SCM for Fruits and Vegetables

Reasons for Returns

- **Spoilage or Damage**: Produce may spoil or get damaged during transportation due to improper temperature control, delays, or inadequate packaging, leading to unusable products upon delivery.
- **Quality Issues**: Customers may return fruits or vegetables if they are overripe, underripe, or show defects such as bruising, mold, or discoloration.
- **Incorrect Orders**: Mistakes in order fulfillment, such as delivering the wrong items or missing products, may lead to returns.
- **Shelf Life Expiry**: Fruits and vegetables are perishable, and if they are stored for too long, they may become unsellable and have to be returned.
- **Consumer Preference**: For direct-to-consumer deliveries, customers might return produce if it does not meet their expectations for taste, appearance, or quality.

Return Logistics

- **Collection and Return Transport**: After a return is initiated, the produce needs to be collected and transported back to a designated location such as a warehouse, processing center, or back to the supplier.
    - **Challenges**: Efficiently transporting returns, especially perishable goods, can be difficult, especially when temperature control is required during the return process.
    - **Reverse Logistics**: The process of returning goods from the customer to the point of origin is known as reverse logistics, and it requires careful management.
- **Handling Returned Products**: Returned produce might not be suitable for resale and could need to be discarded or repurposed.
    - **Composting**: Spoiled produce can be composted to avoid landfill waste, converting it into organic material.
    - **Food Banks or Charities**: Still-edible produce can be donated to food banks, reducing food waste while helping communities in need.
    - **Processing**: Some returned produce, which is slightly damaged but still usable, can be processed into products such as juice, sauces, or other food items.

Using Technology to Manage Returns

- **Tracking Returns**: Technologies like barcode scanners, RFID, and GPS tracking enable efficient monitoring and management of returns, providing visibility into the location and status of returned products.

- **Return Management Systems (RMS)**: Many businesses utilize software that tracks return orders, identifies supply chain issues, and streamlines the return process, improving customer satisfaction and minimizing waste.
- **Blockchain**: Blockchain ensures transparency and traceability in the return process, allowing businesses to identify issues in the supply chain, such as spoilage, and improving overall practices.

Communication with Customers and Stakeholders

- **Clear Return Policies**: Transparent return policies are essential, especially in direct-to-consumer models, so customers understand when returns are accepted and how the process works.
- **Feedback Loops**: Gathering customer feedback on returned produce helps identify recurring problems, improve product quality, and refine quality control processes.
- **Collaboration with Retailers**: Retailers should promptly communicate with suppliers about quality issues or spoilage, enabling quicker resolution and preventing further returns.

Returns' Effect on Sustainability

- **Reducing Food Waste**: Returns, especially with perishable goods, can lead to waste. Sustainable practices like donation, repurposing, or composting can help reduce the environmental impact of returns.
- **Circular Economy**: The circular economy model focuses on reusing, recycling, or repurposing returned goods. For fruits and vegetables, this can involve turning unsellable produce into products like soups, juices, or dried snacks.
- **Packaging Sustainability**: Efficient packaging can reduce damage during transport, decreasing returns due to spoilage or breakage.

Minimizing Returns in the Supply Chain

- **Improved Quality Control**: Regular quality checks at various stages (harvest, packaging, storage, transport) ensure only fresh, high-quality produce reaches consumers and retailers, minimizing returns.
- **Cold Chain Optimization**: Proper temperature control and no delivery delays are vital for maintaining the quality of produce. Investing in better refrigeration systems and monitoring technologies helps reduce spoilage and returns.
- **Demand Forecasting and Inventory Management**: Advanced data analytics can predict demand more accurately, preventing overstocking and ensuring that the right amount of produce is available, reducing returns due to excess stock.

Cost Implications of Returns

- **Financial Impact**: Returns incur costs, including transportation, lost sales, and the need for reprocessing or repackaging returned items.

- **Waste Management Costs**: Disposing of or not repurposing returned produce leads to additional costs, including disposal fees and environmental impact, which can hurt a company's profitability.
- **Customer Loyalty**: Despite the costs, a well-managed return process and clear policies can help retain customer trust, particularly in direct-to-consumer models, promoting long-term loyalty.

Returns in the supply chain management (SCM) of fruits and vegetables present distinct challenges because of the perishable nature of these products. Effectively managing returns demands efficient logistics, transparent communication, technological support, and the adoption of sustainable practices. By prioritizing quality control, minimizing spoilage, and finding ways to repurpose returned goods, businesses can reduce the adverse impact of returns, cut costs, and promote sustainability throughout the supply chain.

TYPES OF SCM MODELS

Supply chain management (SCM) models for fruits and vegetables are primarily centered around strategies that ensure fresh produce is delivered to consumers efficiently while maintaining quality and reducing waste. Below are several common SCM models employed in the fruits and vegetables sector:

Traditional Supply Chain Model

- **Description**: A conventional, linear model where produce flows from farmers to wholesalers, retailers, and ultimately to consumers.
- **Key Characteristics:**
    - Longer lead times due to multiple intermediaries.
    - Limited communication between producers and consumers.
    - Difficulties in maintaining the freshness of perishable goods.
- **Best For**: Small-scale or local producers with limited access to advanced technologies.

Direct-to-Consumer Model

- **Description:** Producers sell directly to consumers, bypassing traditional intermediaries (e.g., via farmers' markets, online platforms, or subscription services).
- **Key Characteristics:**
    - Shorter supply chain, reducing costs.
    - Improved transparency between producers and consumers.
    - Fresher produce as fewer hands handle the goods.
- **Best For**: Local farmers, organic producers, or niche market growers.

Centralized Distribution Model

- **Description**: A central distribution hub aggregates fresh produce from various suppliers, sorts it, and distributes it to retailers or wholesalers.

- **Key Characteristics:**
    - Economies of scale through centralized storage and transportation.
    - Efficient inventory management at the distribution center.
    - Shorter shelf life if not properly managed.
- **Best For**: Large retailers or wholesalers handling bulk volumes.

Cold Chain Model

- **Description**: A temperature-controlled supply chain ensures the freshness and quality of perishable items like fruits and vegetables.
- **Key Characteristics:**
    - Maintains quality, reducing spoilage during transport.
    - Requires refrigerated trucks and storage facilities.
    - Higher costs due to the need for temperature control.
- **Best For**: Long-distance or international trade, especially for refrigerated goods like berries and leafy greens.

Just-in-Time (JIT) Model

- **Description**: This model minimizes inventory by bringing products into the supply chain exactly when needed, reducing waste and overstocking costs.
- **Key Characteristics:**
    - Reduces storage costs and waste.
    - Requires precise demand forecasting and scheduling.
    - Risk of disruptions from sudden demand spikes or delays.
- **Best For**: Retailers or distributors with efficient logistics and accurate demand forecasting.

Collaborative Supply Chain Model

- **Description:** In this model, stakeholders in the supply chain collaborate, sharing information to improve efficiency, reduce costs, and meet demand more effectively.
- **Key Characteristics:**
    - Improved coordination between all parties.
    - Faster responses to market changes or disruptions.
    - Shared risks and rewards in optimizing the supply chain.
- **Best For**: Large-scale operations or industries prioritizing speed and responsiveness.

Agri-Tech/Automated Model

- **Description**: This model uses technology like IoT, AI, and blockchain to enhance supply chain management, tracking freshness, optimizing logistics, and improving inventory management.

- **Key Characteristics:**
    - Increased visibility vía technology across the supply chain.
    - Efficient resource usage through data predictions and route optimization.
    - Real-time tracking and decision-making capabilities.
- **Best For**: Large operations or companies ready to invest in technology for supply chain optimization.

Seasonal and Regional Supply Chains

- **Description:** Focuses on sourcing produce that is locally grown and in season, reducing transportation time and environmental impact.
- **Key Characteristics:**
    - Fresher produce due to shorter transportation times.
    - Lower environmental impact from reduced transportation.
    - Limitations in variety and availability based on season and location.
- **Best For**: Regional retailers and consumers who prioritize fresh, locally sourced produce.

Multi-Tiered Supply Chain Model

- **Description**: Involves multiple layers of suppliers, including local and international sources, with produce passing through several stages before reaching the consumer.
- **Key Characteristics:**
    - Complex supply chain with multiple participants.
    - Potential for delays and logistical issues.
    - Risk diversification due to reliance on multiple sources.
- **Best For**: Large retailers with international supply networks.

10

Supply Chain Logistics and Contract Logistics

Supply chain logistics involves overseeing the flow of goods and services across the entire supply chain, from start to finish. It encompasses the sourcing, transportation, storage, and distribution of raw materials, manufacturing processes, and final products.

The scope of logistics covers the planning and execution of storing and moving goods between different points in the supply chain. It involves coordinating facilities, personnel, equipment, and other resources to ensure that products are transported as planned and have sufficient space at each point in the supply chain.

Key processes such as demand planning, transportation (including fleet management), inventory control, material handling, and order fulfillment all fall under the logistics umbrella. Supply chain logistics is the backbone of global trade, encompassing a wide-reaching network that includes transportation, warehousing, and inventory management systems. This vital aspect of business operations guarantees the efficient movement of products and services from manufacturers to end consumers, passing through various intermediaries along the way.

In today's fast-paced and constantly changing business environment, understanding and optimizing supply chain logistics has become even more critical. Companies are under greater pressure to deliver products faster, more affordably, and with increased transparency.

This section takes a deep dive into the complexities of supply chain logistics, examining the key components, challenges, and how modern technologies play a role in improving these processes. We will look at the strategic decisions made by supply chain executives and the practical applications of inventory management systems, exploring how businesses can enhance their logistics operations to meet consumer demands, control costs, and stay competitive in the global marketplace.

LOGISTICS OF THE SUPPLY CHAIN

Supply chain logistics is a complex system involving numerous organizations working together to get products and services to consumers, making it a crucial component of global trade. This extensive network includes transportation, warehousing, and inventory management, and is often considered the backbone of the global economy.

Logistics executives in the supply chain need to be adept at selecting the most effective transportation modes, designing and establishing warehousing facilities, managing inventory and assets, and creating an efficient logistics network—all while keeping costs low and ensuring exceptional customer service.

Supply chain logistics encompasses the entire process of managing the production, transportation, and distribution of goods within a business's supply chain. This includes both inbound logistics, which deals with how materials and supplies are brought into the company, and outbound logistics, which focuses on how the final products are delivered to retailers and consumers.

Logistics plays a crucial role in ensuring the smooth operation of businesses and the broader global economy, facilitating the flow of goods across borders. Effective logistics management enables businesses to be more resilient in the face of global disruptions and provides a competitive edge in the market.

Logistics management covers a wide range of tasks, including coordinating the sourcing, procurement, and transportation of raw materials to factories; transforming these materials into finished goods; and delivering those goods to retailers and customers. Additionally, it involves managing inventory, warehousing, transportation, distribution, customer service, and returns, all of which are essential for maintaining a smooth supply chain.

Examples of Supply Chain Logistics

Below are the top supply chain management companies in India:

1. **Supply Chain Solutions by TVS:** With the most up-to-date logistics solutions, TVS Supply Chain Solutions is among the top supply chain networks. They offer supply chain management services, such as inventory manipulation, transportation, and warehousing.
2. **Logistics Mahindra:** When it comes to supply chain management and third-party logistics, Mahindra Logistics is an expert. With a "customer first" philosophy, they offer a range of tech-enabled logistics solutions to meet complex needs.
3. **India's DHL Supply Chain:** One of the biggest international express delivery and logistics firms in the world, DHL, has a robust network in India and offers end-to-end logistics services, including transportation, packaging, and warehousing, among other things. They stand out from their competitors thanks to their cutting-edge logistics network.
4. **Logistics for All Cargo:** Allcargo Logistics provides a variety of services, from logistics to global supply chain management, and specializes in multimodal transport operations. They are renowned for their global reach and well-established structure.
5. **Supply Chain Solutions of the Future:** The first fully integrated, IT-enabled end-to-end logistics and supply chain company in India, Future Supply Chain Solutions offers services like supply chain analysis, distribution, and warehousing. They apply contemporary methods to supply chain management.

6. **The Gati Company:** Gati Limited is a business that specializes in providing goods and services as quickly as possible. They are a good partner for businesses all over India because of their extensive network and tech-focused approach.
7. **The Safexpress:** Safexpress provides a full range of logistics services, such as supply chain, transportation, and storage solutions. Their extensive coverage area and effective delivery services are what define them.
8. **Supply Chain Solutions by TCI:** Transportation, warehousing, supply chain engineering, and redesign are all services offered by TCI Supply Chain Solutions. Because their strategy is centered on the needs of the client, each buyer is assured of receiving a customized solution.
9. **The Blue Dart Express:** DHL subsidiary Blue Dart Express Limited is an expert in supply chain management and logistics. They are the go-to option for prompt delivery because of their extensive supply chain and application of contemporary techniques.
10. **DTDC Express Limited:** From supply chain management to individual package delivery, DTDC provides a range of logistics solutions. Their services in the Indian market are driven by their customer service and market awareness. The aforementioned examples highlight how crucial effective SCM procedures are to achieving business success.

These instances show how efficient supply chain management can boost customer happiness, give a company a competitive edge, and spur expansion. The flexibility and significance of supply chain logistics in contemporary business operations are demonstrated by the fact that each company has customized its logistics strategy to fit its unique industry and business model.

How are Supply Chain Management and Logistics the Same?

The movement of goods from the point of origin to the destination is the main focus of both supply chain management and logistics. To ensure that goods can flow through the supply chain as needed, both disciplines necessitate meticulous coordination of labor, facilities, and supplies. Although, it is only one part of the equation, logistics is an essential part of supply chain management.

Similarities of Supply Chain Management and Logistics

Logistics and supply chain management strive for the most efficient transportation, storage, and delivery of goods. The strategic direction that directs inbound and outbound logistics is provided by supply chain management.

Some of the common aspects of SCM and logistics are given below:

- Both are centered around the management and movement of goods, services, or information.
- Both aim to enhance the company's success, differentiating it from its competitors.
- Both prioritize increasing customer satisfaction.

- Both follow a similar process flow of goods and services, starting with the supplier, moving to the manufacturer, then to the wholesaler, and ultimately reaching the retailer or consumer.

How Are Supply Chain Management and Logistics Different?

Supply chain management encompasses the planning, sourcing, production, and delivery of goods, including the management of returns. It defines the overall strategy and activities needed to ensure a smooth flow of products. On the other hand, logistics is concerned with ensuring that the right products arrive at the right location and at the right time, while focusing on the efficient transportation and storage to achieve this.

Differences Between Supply Chain Management and Logistics

As highlighted earlier, supply chain management (SCM) and logistics are closely related but have distinct scopes and focuses.

Key distinctions between SCM and logistics are as follows:

- Logistics are a subset of activities within the broader field of SCM. While logistics deal with the movement and storage of goods, SCM includes additional elements such as production planning, inventory management, labor planning, and the overall delivery of goods and services.
- SCM is primarily concerned with improving processes to gain competitive advantages, whereas logistics focuses on fulfilling customer needs and expectations through the efficient movement of products.
- Logistics concentrates on delivering goods to customers in an efficient and cost-effective manner, while SCM oversees the entire journey of raw materials being transformed into finished products that are moved through suppliers, manufacturers, warehouses, and retailers to consumers.

ROLE OF LOGISTICS IN SUPPLY CHAIN MANAGEMENT

In supply chain management, logistics plays a crucial role in the movement and storage of goods and services, including the documentation and tracking of those movements throughout the entire process until the item reaches the customer.

Logistics encompasses various transportation methods used to move inventory from one point to another. It is responsible for determining optimal storage locations at each stage of the journey, ensuring that goods are readily available when needed for the next phase in the supply chain. Effective logistics are key to the smooth functioning of the overall supply chain management process.

Why Logistics is so Important to Supply Chains?

Logistics plays a vital role in supply chains by overseeing and tracking the people and resources necessary to store and transfer goods and services. It ensures that materials and products are consistently moved at the right time, while staying within budget.

Key aspects of logistics that support supply chains include the following:

- Ensuring the timely delivery of the correct products.
- Reducing costs and enhancing operational efficiency.
- Supporting customer retention and fostering loyalty.
- Offering a distinctive value proposition for certain businesses.
- Facilitating the cost-effective transportation of goods from the production site to the customer's location.

Why is Supply Chain Logistics Important?

Logistics is essential for delivering goods to consumers as quickly and smoothly as possible. Efficient supply chain logistics management helps minimize costs and boost operational efficiency across business functions. By optimizing areas such as demand planning, transportation routes, inventory management, and lead times, companies can save both time and money.

Logistics also plays a key role in maintaining customer satisfaction by ensuring that orders are delivered on time and to the correct location. These advantages contribute to a competitive edge in the market. Effective logistics management enables companies to swiftly respond to market changes, offer better delivery terms, and mitigate disruptions in the most cost-effective manner.

A well-organized supply chain network benefits not only businesses but also consumers and other stakeholders. For customers, efficient logistics results in timely deliveries and accurate order fulfillment, enhancing their shopping experience. Additionally, consumers benefit from effective "reverse logistics"—the process of returning goods to their original location, whether for retail returns, recycling, or disposal.

Well-executed logistics can also be advantageous for retailers and wholesalers. Effective inventory management helps reduce holding costs, while streamlined transportation management lowers shipping expenses, ultimately improving overall profitability.

Logistical Components of the Supply Chain

Supply chain logistics components are essential for managing goods and services throughout the supply chain. Each element plays a vital role in moving materials, finished products, and services through the various stages of the supply chain, ensuring efficient flow from one step to the next.

Below are the logistical components of supply chains:

- **Information:** Information is crucial for tracking the status of items and supply chain processes, helping businesses make informed decisions at every stage.
- **Storage:** Storage involves holding supplies in the correct quantities and locations. It's essential to balance supply and demand to avoid overstocking or running out of stock.
- **Warehousing:** This refers to managing daily warehouse activities, including receiving, storing, picking, packing, and shipping goods.

- **Material Handling:** Material handling pertains to the movement of items within a facility or transport vehicle, and may include the storage, security, and transfer of goods throughout the entire production and distribution process.
- **Packaging:** Packaging ensures that items are delivered undamaged and at the lowest possible shipping cost.
- **Unitization:** Unitization optimizes how goods are arranged, transported, and stored. It also ensures efficient movement using material handling equipment, with the cube being a common form of unitization.
- **Inventory Control:** This component focuses on managing storage and warehousing to optimize stock types, quantities, and locations. Inventory management formulas help calculate demand more effectively.
- **Transportation:** Responsible for moving goods from one stage of the supply chain to the next, transportation involves various modes, such as trucks, trains, ships, planes, and cars, to deliver products to customers.

Challenges of Supply Chain Logistics

Common challenges in supply chain logistics include managing customer service, controlling costs, planning effectively, mitigating risks, and maintaining good supplier/partner relationships. Talent management is also a critical aspect. As businesses expand globally, they face additional challenges in being flexible enough to grow into new markets while staying competitive.

Supply chain logistics executives today oversee multiple chains and strive to meet the needs and expectations of both customers and suppliers. The rise of personalized offerings adds complexity, as managing this personalization presents a significant logistics challenge.

To address these challenges, companies are leveraging advanced systems like supply chain management tools, customer relationship management platforms, and Big Data. These technologies help businesses gain valuable insights into their customers, making logistics operations more efficient, cost-effective, and able to meet customer demands.

UNDERSTANDING CONTRACT LOGISTICS

Contract logistics refers to the practice of outsourcing a company's supply chain and logistics operations to a third-party provider. This provider, known as a contract logistics specialist, manages the various aspects of the logistics process on behalf of the company.

Contract logistics involves a long-term partnership where a logistics service provider offers a comprehensive range of services, from transporting products or spare parts to delivering them to the final consumer. This requires the provider to own and manage necessary resources such as warehouses, equipment, personnel, stock management systems, dispatch monitoring, and after-sales services.

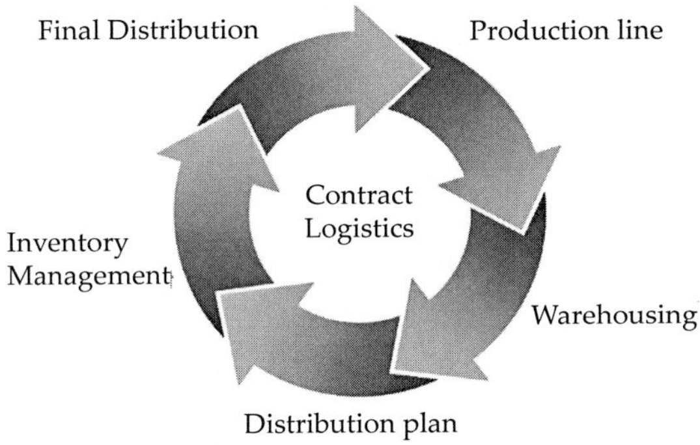

Figure: 10.1 Contract logistics

In this arrangement, the logistics provider becomes a crucial part of the supply chain, connecting the various partners of the industrial company or retailer to ensure smooth operations. Contract logistics is especially suited for businesses engaged in long-term service agreements. The provider handles every phase of the distribution process, from initial stages to final delivery.

Details of Contract Logistics

Logistical solutions are designed to cater to the needs of various companies, ranging from storage to distribution, both domestically in France and internationally. These solutions are customized to align with the specific product types, company values, and their target customer base. The offered services include:

- **Storage**: Managing appropriate facilities for storing goods.
- **Fulfillment (Order Processing)**: Handling the processing of orders from receipt to delivery.
- **Added Value Services**: Offering services like stock management optimization to improve efficiency.
- **Upstream, Downstream, and On-Site Logistics**: Managing procurement, packaging, and availability of spare parts on-site.
- **Spare Part Management**: Managing the storage, stock, and efficient use of spare parts.
- **Foreign-Trade Zone (FTZ)**: A regulated area where goods can be delivered, stored, processed, and exported, adhering to specific customs regulations.

Examples and Practical Applications

The logistics provider managing your supply chain should be strategically located in France or abroad, depending on your market and its specific requirements. Opting for a market leader with a strong reputation in France is recommended:

- **Comprehensive Management**: They must be equipped to address all logistical challenges using highly skilled human resources. From production through to storage and distribution, they should be able to adapt operations to both industrial and retail contexts.

- **Tailored Analysis**: The provider should begin by analyzing your requirements and identifying your specific logistical needs. The aim is to enhance your company's growth by offering high-value services that give you a competitive edge and improve customer satisfaction.

A key example is after-sales service, which can be complex in reverse supply chains. By including this as part of contract logistics, significant added value is created:

- The logistics provider manages after-sales services and product returns, ensuring timely collection from customers.
- They ensure products are reconditioned for their second life cycle.
- The provider may even handle disassembly and reassembly for products at the end of their life.
- For companies focused on sustainable development, additional services are included in contract logistics to efficiently manage eco-friendly practices.

Another service logistics providers should offer is collecting and recycling returned goods. Items that are beyond repair are disassembled, and in compliance with legal requirements, they are recovered, recycled, or destroyed. The value of these components is factored into your company's financial results, and the reuse of certain parts can contribute further to sustainability efforts.

Bibliography

Acharya, S. S. & Agarwal, N. L. (2012). Agricultural Marketing in India. Oxford & IBH Publishing.

Agrawal, D. K., & Patil, R. K. (2017). "Cold Chain Logistics in India: Issues and Innovations." Journal of Agricultural Economics and Development, 5(2), 32-45.

Amul Dairy. (2020). Innovations in Dairy Cold Chain and Processing in India.

Amul Dairy. (2021). Cold Chain Management for India's Dairy Sector: Case Study on Amul.

APEDA. (2021). Advancements in Cold Storage and Logistics for Export-Oriented Agro-Processing Industries.

APEDA. (2021). Cold Chain Development in India: Challenges and Opportunities. Government of India.

Asian Development Bank. (2018). Enhancing Agricultural Supply Chains in India through Cold Chain Development.

Bhattacharya, S. (2014). Postharvest Technology of Fruits and Vegetables: General Concepts and Principles. CBS Publishers.

Britannia Industries. (2019). Supply Chain Management for Perishable Food Products in India.

Chaudhary, R., & Sharma, A. (2019). "Role of Food Processing Industry in Enhancing Farmer's Income in India." Indian Journal of Economics and Development, 15(3), 78-92.

CII. (2019). Strengthening India's Agricultural Value Chain: Cold Chain and Warehousing Solutions.

CII. (2020). Food Processing and Logistics: Enhancing Value Chain Efficiency in India. Confederation of Indian Industry.

CII. (2021). Enhancing Agro-Processing Sector Growth through Cold Chain Innovations.

CSIR-Central Food Technological Research Institute. (2019). Technological Interventions in Cold Storage and Value Addition in India.

CSIR-CFTRI. (2020). Food Processing & Preservation: Innovations for Indian Agriculture.

Deloitte India. (2019). Smart Cold Chain Management for India's Food Industry.

Deloitte India. (2020). Smart Logistics and Cold Chain Management for the Indian Food Industry.

Department of Commerce, Government of India. (2021). Export Potential for Indian Agro-Processed Products: Enhancing Cold Chain Infrastructure.

FAO. (2021). Cold Chain Development in India: Lessons from Global Best Practices.

FICCI. (2018). Cold Storage and Refrigeration: A Roadmap for India's Food Sector.

FICCI. (2019). Building Efficient Cold Chain Infrastructure in India: Policy and Market Insights. New Delhi.

FICCI. (2021). Role of Logistics in Enhancing Food Processing Sector Growth in India.

Food Processing Journal India. (2020). Advancements in Value-Added Products from Indian Agriculture.

Food Safety and Standards Authority of India (FSSAI). (2018). Guidelines for Cold Storage and Food Safety Standards. New Delhi.

Food Safety and Standards Authority of India (FSSAI). (2019). Regulatory Framework for Cold Chain Infrastructure in India.

Goyal, S. K. & Joshi, P. K. (2018). Agricultural Value Chains in India: Prospects and Challenges. Academic Foundation.

Gujarat Agro Industries Corporation (GAIC). (2020). Cold Chain and Agro Processing in Gujarat: Investment and Growth Opportunities.

Gupta, R. & Mehta, A. (2020). "Value Addition in Agricultural Produce: Trends and Challenges in India." Journal of Rural Development Studies, 12(4), 45-59.

Haldiram's. (2020). Frozen Foods and Cold Storage Management in India.

ICAR-CIPHET. (2020). Advances in Food Processing Technologies in India.

ICRISAT. (2021). Postharvest Technologies for Pulses and Cereals in India.

IFC. (2021). Investment Opportunities in Cold Chain Infrastructure in India.

IMARC Group. (2021). India Cold Chain Market Report: Growth, Trends, and Forecast 2021-2026.

Indian Cold Chain Development Forum (2021). Revolutionizing the Indian Cold Chain Sector: Policies & Market Growth.

Indian Council of Agricultural Research (ICAR). (2018). Agro-Processing Industries and Cold Storage Development in India.

ICAR-CIPHET. (2020). Cold Chain Management for Perishable Products. Technical Bulletin, Ludhiana, India.

Kadam, D. M., and Kumar, R. (2012). Development of Value-Added Products from Underutilized Fruits. Indian Journal of Agricultural Sciences, 82(8), 715–719.

Kader, A. A. (2002). Postharvest Technology of Horticultural Crops (3rd edition). University of California, Agriculture and Natural Resources Publication 3311, Davis, California.

Kitinoja, L., and Kader, A. A. (2002). Small-Scale Postharvest Handling Practices: A Manual for Horticultural Crops. Postharvest Horticulture Series No. 8, University of California, Davis.

Ministry of Food Processing Industries (MOFPI). (2020). Guidelines for Infrastructure Development and Mega Food Parks. Government of India. Available at: www.mofpi.nic.in

Mittal, J. P. (2007). Agricultural Processing and Food Engineering. Jain Brothers, New Delhi.

National Centre for Cold-chain Development (NCCD). (2016). All India Cold-chain Infrastructure Capacity Assessment. Ministry of Agriculture, Government of India. Available at: www.nccd.gov.in

Panigrahi, S., and Prasad, S. (2015). Value Addition in Agricultural Products: Opportunities and Challenges. Agrotech Publishing Academy, Udaipur.

Parry, J. L. (1985). Packaging and Storage of Fruits and Vegetables. FAO Agricultural Services Bulletin No. 59, Rome.

Ranganna, S. (2001). Handbook of Analysis and Quality Control for Fruit and Vegetable Products (2nd edition). Tata McGraw-Hill, New Delhi.

Rathore, D. S., and Bhargava, B. S. (2008). Fruit Processing. Agrobios India, Jodhpur.

Sanwal, S. K., and Singh, B. (2013). Postharvest Management and Value Addition in Vegetables. Directorate of Knowledge Management in Agriculture, ICAR, New Delhi.

Sharma, G. P., and Prasad, S. (2004). Effective moisture diffusivity of onion slices during infrared drying. International Journal of Food Properties, 7(3), 603–615.

Singh, D. B., and Jain, R. K. (2010). Engineering Properties and Processing of Agricultural Products. Saroj Prakashan, Allahabad.

Singh, K. K., and Sahay, K. M. (2001). Unit Operations of Agricultural Processing (2nd edition). Vikas Publishing House Pvt. Ltd., New Delhi.

Verma, L. R., and Joshi, V. K. (2000). Postharvest Technology of Fruits and Vegetables: Handling, Processing, Fermentation and Waste Management. Indus Publishing Company, New Delhi.

Index